BIOLOGICAL AND MEDICAL PHYSICS,
BIOMEDICAL ENGINEERING

BIOLOGICAL AND MEDICAL PHYSICS, BIOMEDICAL ENGINEERING

The fields of biological and medical physics and biomedical engineering are broad, multidisciplinary and dynamic. They lie at the crossroads of frontier research in physics, biology, chemistry, and medicine. The Biological and Medical Physics, Biomedical Engineering Series is intended to be comprehensive, covering a broad range of topics important to the study of the physical, chemical and biological sciences. Its goal is to provide scientists and engineers with textbooks, monographs, and reference works to address the growing need for information.

Books in the series emphasize established and emergent areas of science including molecular, membrane, and mathematical biophysics; photosynthetic energy harvesting and conversion; information processing; physical principles of genetics; sensory communications; automata networks, neural networks, and cellular automata. Equally important will be coverage of applied aspects of biological and medical physics and biomedical engineering such as molecular electronic components and devices, biosensors, medicine, imaging, physical principles of renewable energy production, advanced prostheses, and environmental control and engineering.

M.I. Monastyrsky (Ed.)

Topology in Molecular Biology

With 118 Figures, 6 in Color and 3 Tables

Springer

Professor Dr. Michail Ilych Monastyrsky
Institute of Theoretical and Experimental Physics
B. Cheremushkinskaya 25, 117259 Moscow, Russia
E-mail: monastyrsky@itep.ru

ISSN 1618-7210
ISBN-13 978-3-642-06242-1 e-ISBN-13 978-3-540-49858-2

Springer is a part of Springer Science+Business Media

springer.com

Cover concept by eStudio Calamar Steinen

Foreword

The contents of this book focus on the recent investigations in molecular biology where applications of topology seem to be very stimulating. The volume is based on the talks and lectures given by participants of the three-month program "Topology in Condensed Matter", which was held in the Max Planck Institut fur Physik komplexer Systeme, Dresden, Germany, 8 May–31 July 2002, under the scientific direction of Professors M. Kléman, S. Novikov and myself. The aim of this program was to discuss recent applications of topology to several areas in condensed matter physics and molecular biology.

The first volume "Topology in Condensed Matter" is concerned with modern applications of geometrical and topological techniques to such new and classic fields of physics like electron theory of metals, theory of nano-crystals, aperiodic and liquid crystals, quantum computation and so on. This volume is published simultaneously in "Springer Series in Solid-State Physics".

The present volume gives an exposition of the role of topology in the theory of proteins and DNA. The last thirty years affirmed very efficient applications of modern mathematics, especially topology, in physics. The union of mathematics and physics was very stimulating for both sides. On the other hand, the impact of mathematics in biology has been rather limited. However here also some interesting results were obtained. In particular, there are applications of knot theory in the theory of circular closed DNA. The recent discoveries in molecular biology indicate future successful applications of topology. For example, a reconstruction of three-dimensional protein structures by one-dimensional genomic sequences leads to very interesting and non-trivial combinatoric problems. There exist two "principa" reflecting the state of affairs in both fields: physics and biology in the recent past. The first one is the very popular concept of the famous physicist E. Wigner about "the unreasonable effectiveness of applications of mathematics in natural sciences (i.e. physics)". Otherwise there exists the opposite opinion of the renowned contemporary mathematician I. Gelfand, who worked for many years in mathematical biology. He expressed the "unreasonable non-effectiveness of applications of mathematics in biology". It is not to say that there are no applications of mathematics to biology, but in Gelfand's view, no in-depth applications. May be the future development of both disciplines will disprove this joke. One indirect proof of this tendency is the contribution of Gelfand himself in this volume. Beside the biological chapter we add a mini-course of topology for physicists and biologists. We hope that this mathematical supplement makes this book self-contained and comprehensible for a more broad audience, including graduate and undergraduate students. Our biology chapter contains accounts of the recent interactions of topology and molecular biology – interactions with indeed much depth.

By the common opinion of participants the seminar was very successful. The organizers and participants are grateful to the MPIPKS for the generous sponsorship of the seminar with so unusual spectra of interest. Special thanks go to the directors of, MPIPKS Professors P. Fulde, J.-M. Rost and F. Julicher, the head of visitors' program Dr. S. Flach, the secretaries K. Lantch, M. Lochar and C. Poenish. We acknowledge our gratitude to the entire staff of the Institute for their help in organizing the seminar and for making sure it ran smoothly. We acknowledge our gratitude to Dr. C. Ascheron, who suggested publishing these lectures in Springer Verlag, and Sabrina Gauthamee Khan and K. Venkatasubramanian of SPi, Pondicherry and Adelheid Duhm who assisted in preparation of these books. The editor especially thanks Dr. L. Alania for his assistance in preparing this volume. We hope such programs that converge mathematicians, physicists, and biologists will continue.

Moscow-Dresden, November 2005. *Michael Monastyrsky*

Contents

List of Contributors

R. Brooks
Department of Mathematics
Technion — Israel Institute
of Technology
Haifa
Israel
http://www.math.technion.ac.il/~rbrooks

V.M. Buchstaber
Steklov Mathematical Institute
117966, Moscow
Russia
buchstab@mendeleevo.ru

C. Cerf
Département de Mathématique
Université Libre de Bruxelles
B-1050 Bruxelles
Belgium
ccerf@ulb.ac.be

P. Favorov
Engelhardt Institute of Molecular
Biology,
Russian Academy of Sciences
32 Vavilov St., 117984
Moscow, Russia

A. Gabibov
Shemyakin and Ovchinnikov
Institute of Bioorganic Chemistry
Russian Academy of Sciences
16/10 Mikluho-Maklaya St.
117871, Moscow
Russia

I.M Gelfand
Department of Mathematics
Rutgers University
Piscataway, NJ 08854
USA

T.I. Gelfand
Department of Mathematics
Rutgers University
Piscataway, NJ 08854
USA

L.H. Kauffman
Department of Mathematics,
Statistics and Computer Science
University of Illinois at Chicago
851 South Morgan St.
Chicago IL 60607-7045
USA
kauffman@math.uic.edu

A.E. Kister
Department of Mathematics
Rutgers University
Piscataway, NJ 08854
USA

M.V. Kleyzit
Department of Mathematics
Rutgers University
Piscataway, NJ 08854
USA

S. Lambropoulou
Department of Mathematics
National Technical University
of Athens
Zografou campus
GR-157 80 Athens
Greece
sofia@math.ntua.gr

S.D. Levene
Department of Molecular
and Cell Biology
and Institute of Biomedical Sciences
and Technology
University of Texas at Dallas
PO Box 830688, Richardson
TX 75083-0688
USA

M. Lukin
Institute of Experimental Cardiology,
Cardiology Research Center,
15A 3-d Cherepkovskaya St.
121552, Moscow
Russia

D.V. Millionschikov
Department of Mathematics
and Mechanics
Moscow State University
119899, Moscow
Russia
million@mech.math.msu.su

M. Monastyrsky
Institute of Theoretical
and Experimental Physics
117259, Moscow
Russia
monastyrsky@itep.ru

A. Reshetnyak
Chemical Department
Moscow State University
Vorobjovy Gory, 119899
Moscow, Russia

N. Rivier
ACI biofrustration
Laboratoire de Dynamique des
Fluides Complexes
Université Louis Pasteur
67084 Strasbourg
France
nicolas.rivier@fresnel.
u-strasbg.fr

J.-F. Sadoc
ACI biofrustration
Laboratoire de Physique
des Solides -
Université Paris Sud
(associé au CNRS), Bât. 510
91405 Orsay, France

A. Stasiak
Laboratoire d'Analyse
Ultrastructurale, Bâtiment de
Biologie
Université de Lausanne
CH-1015 Lausanne-Dorigny
Switzerland
Andrzej.Stasiak@lau.unil.ch

A. Vologodskii
New York University
New York, NY 10003
USA
alex.vologodskii@nyu.edu
Telephone: 212-998-3599
Fax: 212-260-7905

E. Yakubovskaya
Shemyakin and Ovchinnikov
Institute of Bioorganic Chemistry
Russian Academy of Sciences,
16/10 Mikluha-Maklaya St.
117871, Moscow
Russia

1

Introduction

M. Monastyrsky

The problems mostly discussed in this volume pertain to the relationship between 1D and 3D structural date in proteins and DNA. This theme began in the last decades of the XX century, and is still the focus of numerous biophysical discussions. For years, the main question of how linear genome sequences predetermine the spatial structure of bipolymer remained very intriguing. The recent genome sequence analysis has provided new tools for studying DNA and proteins.

These problems lead, besides new biological questions, to an interesting mathematics, which is very natural to topology and more precisely to knot theory. Most of the chapters presented in this volume are concerned with these topics.

The book begins with a chapter by S.D. Levene, which might be considered as an introduction to the topological aspects of DNA structures. In the next chapter "Monte Carlo simulation of DNA Topological properties" Vologodskii studies the problem of calculating the main quantity writhing. In the chapter "Dynamics of DNA supercoiling". Gabibov et al. are concerned with the very interesting and recently studied dynamics of supercoiled DNA, with topological constraints. The authors analyse not only theoretical aspects of the problem but also the experimental situation.

The following two chapters of Kauffman and Lambropoulu "From Tangle Fractions to DNA" and Cerf and Stasiak "Linear behavior of the writhe versus the number of crossings in rational knots and links" are devoted to interesting topological problems related with recombination properties of DNA. We point out that it is a rare case where biological questions lead to new mathematical notions such as the theory of tangle equations.

The next section commerces with the chapter of Kister et al. The authors provide the combinatorical analysis of the above-mentioned problem: how one-dimensional genomic sequences determine three-dimensional protein structure. One more special problem is the structure of collagen, which is a protein with periodic structure. Rivier and Sadoc study the assembly of collagen molecules, the so-called fibrils, long, periodic bundle of finite collagen molecules. The appearance of three-dimensional periodic structures leads to very interesting geometrical questions similar to the problems of classification textures and defects in liquid crystals (smectics and discotics), lattices of defects in superconductors, defects in liquid membranes, dense packing of spheres and so on.

The book ends with a large mathematical supplement. A short course on topology is included, assuming that some knowledge of topology presented in a comprehensive form will be useful for physicists and biologists. The basic notions in topology already used in biology as a reader can be found in the biological chapters of the book and some background is given. We guess it will be useful in future investigations.

The lectures of Buchstaber and Monastyrsky can also be found in this compendium. The points raised in the lecture of Millionschikov based on the recent new developments of topology, the theory of multivalued functionals, have already been applied in physics. It seems that such a good technique will be useful in future applications in biology. I follow the thought-provoking motto of John von Neumann: "Modern mathematics can be applied after all. It is not clear a priori, is it, that could be so".

The last chapter written by Brooks is based on his colloqium lecture in the MPIPKS, Dresden, and also on his talk in the Institute d'Henri Poincare, Paris. He considered some relations between graph theory and spectral properties of Laplacians on Riemann surfaces.

We publish his lecture for two purposes. First of all it is a very good mathematic study concerning with two fundamental topics (graph theory and Riemann surfaces) with very promising applications in biology. The second one is to acquaint a more general audience with the work of Robert Brooks, the very deep and original mathematician. He participated very actively in our seminar. Unfortunately he passed away soon after the end of our program. We dedicate this volume to his memory.

2

Topology in Biology: From DNA Mechanics to Enzymology

S.D. Levene

Summary. The focus of this contribution is on biological applications of topology to the study of DNA structure and to understanding protein–DNA interactions that involve alterations of DNA topology.

We review basic aspects of DNA structure and the tertiary organization of circular DNA by supercoiling, knotting, and catenation (linking). This is followed by a review of our current understanding of the topology of chromosomal DNA. Finally, we discuss some topological and structural aspects of DNA site-specific recombination by the yeast enzyme FLP.

2.1 Overview

The focus of this contribution is on biological applications of topology to the study of DNA structure and to understanding protein–DNA interactions that involve alterations of DNA topology. Control of DNA topology is an essential aspect of the existence of every living cell because of the extraordinary degree to which the genomes of free-living organisms are confined. Moreover, changes in DNA topology accompany a wide range of enzyme-mediated processes on DNA such as replication, recombination, and repair. In higher organisms the interconversion of DNA between an inert state contained within chromatin fibers and an active state characterized by greater, but not necessarily complete, accessibility, provokes many questions about how DNA topology and its regulation provides both challenges and opportunities to the cell.

We first review basic aspects of DNA structure and the tertiary organization of circular DNA by supercoiling, knotting, and catenation (linking). This is followed by a review of our current understanding of the topology of chromosomal DNA. Finally, we discuss some topological and structural aspects of DNA site-specific recombination by the yeast enzyme FLP.

2.1.1 Why Study DNA Topology?

Topological aspects of DNA structure can provide great insight into biochemical mechanisms of proteins that mediate changes in DNA structure and topology. The goal of these studies is generally to understand how the structure of a DNA-metabolizing enzyme and that of the DNA sequence recognized by the enzyme interact to participate in a particular chemical reaction. The overall change in DNA topology that takes place often greatly limits the number of prospective mechanistic scenarios because any changes in topology must be consistent with overall changes in DNA geometry.

A limitation of this approach is that although DNA topology and geometry must be consistent with one another, the latter is rarely uniquely determined. However, there is at least one important advantage of the topological approach that outweighs any of its disadvantages: the fact that the topology of a DNA molecule is fixed and invariant as long as the backbones of both DNA strands remain unbroken. As long as this constraint is not violated, perturbations of DNA structure do not affect its global topology. The topological state of a DNA molecule is independent of temperature, solution conditions, the presence of particular ions or small DNA-binding molecules, or any other environmental factors, which offers an enormous experimental advantage.

Finally, an underappreciated aspect of the topological approach is that topology is extremely useful in ruling out implausible mechanistic scenarios. One is sometimes faced with the prospect of selecting the most likely mechanism from a long list of candidates. Frequently, the availability of topological information helps to limit the plausible choices to a small subset or a unique scheme.

2.1.2 Secondary and Tertiary Structure of DNA

The helical structure of double-stranded DNA is an integral aspect of the topology of closed DNA molecules. Figure 2.1 shows the three canonical structures of double-stranded DNA one frequently encounters in textbooks (see [1] for example). Over the last 20 years it has become clear that local sequence-dependent variations on these canonical themes exist; therefore, these structures should be thought of as prototypes of structural families rather than rigid templates.

The B form of DNA is the structure considered most representative of DNA molecules in aqueous solution: it is a right-handed double helix with a period of 10.5 residues (base pairs). Each base pair is nearly perpendicular to the helix axis and separated from its neighbors by 0.34 nm, giving the helix a pitch of 3.6 nm. The A-form DNA is a particular structural family characteristic of DNA molecules under conditions of poor hydration (DNA fibers at low relative humidity or molecules in solution that contain substantial amounts of alcohol or other nonaqueous solvents). It is also a right-handed helical form with a period of 11.0 base pairs and a pitch of 3.2 nm. Unlike the

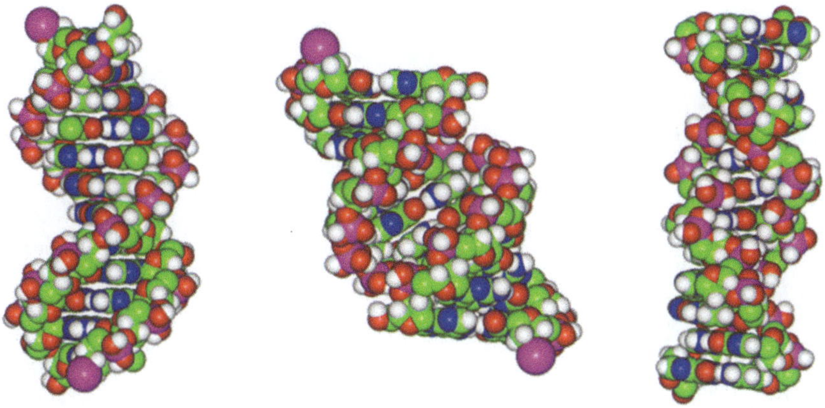

Fig. 2.1. Canonical structures of DNA generated according to sequence-averaged values of helical parameters in the corresponding geometries

B form in which the centers of the base pairs are nearly coincident with the helix axis, the base pairs in A-form DNA are displaced from the central axis. This feature, along with the strong inclination of A-form base pairs relative to the helix axis, imparts radically different geometries to the B and A forms. The Z form of DNA is a structure that is particular to alternating purine–pyrimidine sequences such as GC and is present only under conditions of high ionic strength in vitro or when DNA is negatively supercoiled (underwound). It is a left-handed helical structure with 12.0 base pairs per turn and a pitch of 4.5 nm. The biological significance of Z-form DNA has been a matter of significant controversy [2, 3].

Subtle details of local DNA structure and/or the juxtaposition of these different structural forms along the same DNA molecule can direct the global structure of a particular DNA sequence [4, 5]. This is largely because the thermodynamic stability of DNA structures depends on the favorable stacking of base pairs in the interior of the double helix and attendant exclusion of water from the hydrophobic interior of the helical structure. Because individual base pairs can vary substantially in terms of their inclination relative to the helix axis, the tendency of base pairs to stack generates local deflections in the DNA-helix axis. An example of this phenomenon is shown in Fig. 2.2, in which alternating helical structures generate a series of small bends in the double helix. When repeated periodically in phase with the helix screw, such patterns can generate large-scale intrinsic bends in DNA [6].

2.1.3 DNA Flexibility

For DNA molecules in solution (and presumably also in the cell), the influence of local DNA structure is attenuated by thermal Brownian motion.

Fig. 2.2. Structure of an intrinsically bent DNA sequence in which alternating tracts of $[dA{:}dT]_5$ and random-sequence DNA are arranged in phase with the DNA helical screw. There are ten $[dA{:}dT]_5$, which each generate an average intrinsic bend of 18 degrees [45, 46]. Given the near-perfect phasing of these intrinsic bends, the overall intrinsic bend generated over the entire 105 base-pair sequence is approximately 180 degrees

The polyelectrolyte nature of DNA molecules confers a high degree of bending rigidity on the double helix; however, double-stranded molecules that are any larger than about 30 base pairs deviate significantly from rigid-rod behavior. It is more useful to consider the superposition of modes of thermal flexibility, which may be isotropic or anisotropic, on the sequence-dependent intrinsic structure of DNA molecules. The theoretical description of semiflexible polymer chains according to the wormlike-chain model [7] provides a nearly ideal framework for analyzing DNA tertiary structure.

The wormlike-chain model postulates a resistance to local bending that is proportional to the angular deviation from the chain's equilibrium conformation. The contribution to the total energy of a wormlike chain from a chain segment i,u_i, is given by

$$u_i = \alpha_i \left(\theta_i - \theta_i^0\right)^2 , \tag{2.1}$$

where θ_i is the angular displacement of the ith segment relative to segment $\mathrm{i}-1$, θ_i^0 is the value of θ_i in the chain's minimum-energy conformation, and α_i is a bending energy constant for this displacement. This expression is recognizable as a classical Hooke's law potential for the local deformation of an elastic rod.

Intrinsically straight wormlike chains, those whose minimum-energy conformations are that of a perfectly straight rod, have tractable closed-form expressions for the mean-square end-to-end distance, $\langle \mathbf{h}^2 \rangle$,

$$\langle \mathbf{h}^2 \rangle = \langle \mathbf{h} \cdot \mathbf{h} \rangle = 2P \left[L - P \left(1 - \mathrm{e}^{-L/P} \right) \right], \tag{2.2}$$

where $\mathbf{h}$ is the end-to-end vector of the polymer chain, L is its contour length, and P is a parameter called *the persistence length.* P is a quantity with dimensions of length that characterizes the bending rigidity of the wormlike chain model and measures the tendency of an intrinsically straight chain to propagate in the direction specified by the tangent to the chain at one end. It is useful to consider the asymptotic behavior of the expression for $\langle \mathbf{h}^2 \rangle$ in the limit of both small and large L, namely

$$\lim_{L \to 0} \frac{\langle \mathbf{h}^2 \rangle}{L^2} = 1 \quad \text{and} \quad \lim_{L \to \infty} \frac{\langle \mathbf{h}^2 \rangle}{L^2} = \frac{1}{L}. \tag{2.3}$$

Thus, in the limit of small wormlike chains, the polymer's end-to-end vector tracks the contour whereas the displacement of the ends of large chains grows with an $L^{1/2}$ dependence, identical to that of a nonself-avoiding random walk. P is a kind of correlation length for the chain, which can be appreciated from the formula for the average component of the end-to-end vector in the initial chain direction, $\langle \mathbf{h}_z \rangle$

$$\langle \mathbf{h}_z \rangle = \langle \mathbf{h} \cdot \hat{\mathbf{z}} \rangle = \frac{1}{X} \left(1 - \mathrm{e}^{-X} \right), \tag{2.4}$$

where $X = L/P$. For a wormlike chain with $L = P$ (equal to about 50 nm or 150 base pairs for DNA under physiological conditions), the component of the end-to-end vector in the chain's initial direction is only 63% of the overall contour length. The fact that chains of this length do not behave as rigid rods is shown in Fig. 2.3, which shows the extent of conformational space occupied by several Boltzmann-sampled chain conformations propagating initially in the same ($z-$) direction.

The concept of persistence length as a measure of intrinsic bending flexibility makes sense only for wormlike chains whose minimum-elastic-energy conformation is that of a straight rod. Wormlike chains that have distorted elastic minima deviate from a rodlike structure even in the absence of thermal fluctuations and thus have a structural or static component to the persistence length. Thermal fluctuations, the magnitude of which depend on the local bending modulus of the chain and the absolute temperature, are superposed on this minimum-energy shape. The value of the bending energy constant in (2.1), α_i, can be related to the persistence length of an intrinsically straight molecule with the same flexural elasticity, the so-called dynamic persistence length, P_d

$$\alpha_i = \alpha = 2k_\mathrm{B}TP_\mathrm{d}/\ell, \tag{2.5}$$

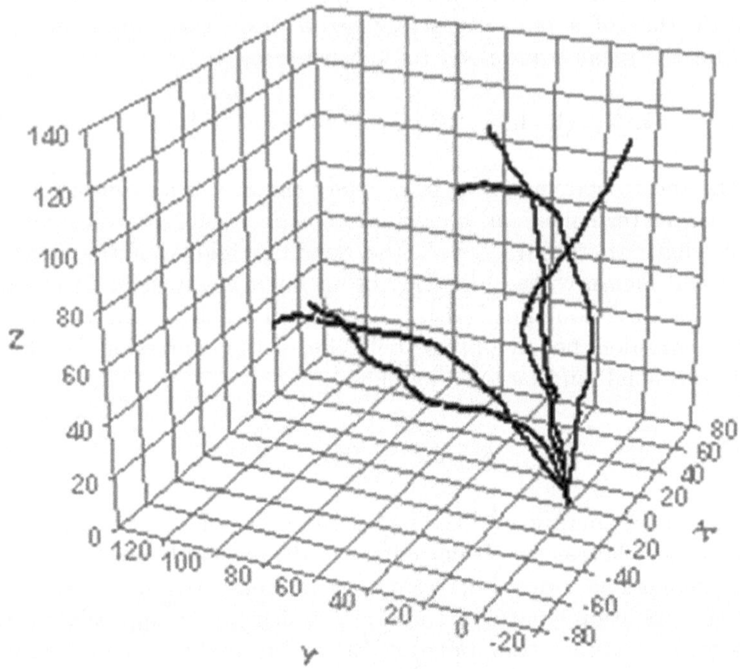

Fig. 2.3. Several conformations of an intrinsically straight wormlike chain with contour length, L, equal to the persistence length, P. Conformations were simulated by a Monte Carlo computer algorithm; each chain consisted of 150 rigid segments connected to its neighbors by a semiflexible joint. Chains were fixed at a common origin and assigned identical initial directions of propagation

where P is the physical length of a segment in the wormlike chain model, T is the absolute temperature, and k_B is Boltzmann's constant. We assume here that P_d is independent of the DNA sequence and that bending flexibility is isotropic; however, more sophisticated treatments that relax these conditions are sometimes warranted.

Many techniques have been used to measure the persistence length of DNA and other nucleic-acid structures, including rotational and translational diffusion [8, 9], ligase ring-closure kinetics [10, 11], and electron and atomic-force microscopy [12, 13]. Discussion of these methods is beyond the scope of this contribution; the reader may wish to consult a recent review [14] or any of the above citations for details. Despite disagreements on the value of P in the early literature, recent measurements have converged on a value of about 50 nm at moderate ionic strength. As discussed above, this value superposes the effects of intrinsic flexibility, as manifested in the dynamic persistence length, and nonuniformity of DNA structure, characterized by the static value of the persistence length, P_s. For DNA fragments that represent a random assortment of DNA sequence elements, Schellman and Harvey [15] derived an expression that relates the measured value P to its component dynamic and static contributions

$$\frac{1}{P} = \frac{1}{P_d} + \frac{1}{P_s}. \tag{2.6}$$

2.1.4 Topology of Circular DNA Molecules

The topological organization of double-stranded DNA is intimately connected with the biology of the molecule. Most bacterial cells have circular genomes consisting of a covalently closed double-stranded chromosome. Because the two DNA strands are linked, scission of at least one, but usually both, of the strands is essential for segregation of daughter chromosomes during cell division. In addition, many aspects of DNA metabolism such as DNA synthesis, transcription, and recombination generate torsional stress that leads to underwinding or overwinding of the double helix. This torsional stress partitions into local changes in the twist number of DNA (number of base pairs per helix turn) and also global winding of the DNA helix axis, termed *writhe*. Although the genomes of eukaryotic cells consist of linear DNA molecules, similar topological constraints apply due to the binding of architectural proteins that maintain a scaffold structure within chromosomes (see below). In eukaryotic genomes overall organization is highly complex with multiple levels of DNA winding mediated by histone proteins, nucleosomal association, and chromatin condensation. However, with approximately one gene located in every 50,000 base pairs in the human genome [16], it is not unreasonable to expect that activating accessibility of genes to the cell's transcriptosome, which involves remodeling of chromatin structure, would require maintaining topologically independent domains every 50,000 base pairs, on average.

The topology of a covalently closed DNA molecule is described in terms of a mathematical quantity called *the linking number*, Lk. Formally, Lk is one-half the sum of signed crossings of the DNA single strands (see Fig. 2.4). Lk is a topological invariant; no distortion of DNA structure short of breaking one or both DNA strands alters Lk. Based on the work of Calugareanu [17], White [18] and Pohl [19] showed that Lk is related to the local and global geometry of a pair of linked space curves through the formula

$$\mathrm{Lk} = \mathrm{Tw} + \mathrm{Wr}, \tag{2.7}$$

where Tw is the total twist of the space curves about the central axis and Wr is the self-linking number or writhe of the central axis. Because Lk is a topological invariant, thermal fluctuations in Tw and Wr occur subject to the constraint in (2.7). Relations between the geometry of a particular DNA conformation and Tw and Wr, the latter in terms of the famous Gauss integral, can be found in [20].

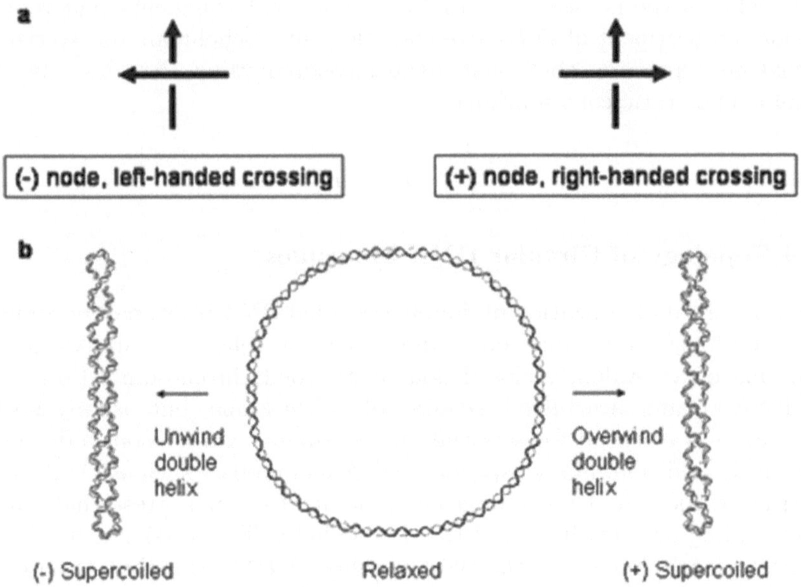

Fig. 2.4. Topology of closed circular DNA. (**a**) Sign convention for DNA crossings in closed-circular DNA. The convention corresponds to the normal right-hand rule in chemistry and physics: a left-handed crossing is counted as negative whereas a right-handed crossing is counted as positive. (**b**) Conversion of relaxed DNA into negatively and positively supercoiled DNA. The descriptor of supercoiling, the linking number, can be computed from one-half the sum of signed crossings of the black and red strands. In the case of relaxed DNA there is no writhe and the linking number, Lk, is equal to the twist number, Tw. In negatively supercoiled DNA, reduction of Lk below Tw gives rise to right-handed interwound supercoils, or negative writhe. Conversely, incrementing Lk above Tw generates left-handed interwound supercoils and positive writhe

Almost all DNA in the cells of terrestrial organisms is underwound or negatively supercoiled, which means that the Lk of DNA molecules is reduced below the value that pertains to the same DNA in the absence of torsional stress, designated Lk_0. Exceptions to this rule are found among archeabacteria that require positively supercoiled genomes in order to survive in extreme conditions of temperature and pressure near geothermal vents at the ocean floor. The distortion of DNA structure generated by negative (deficits in Lk relative to Lk_0, $\Delta Lk < 0$) or positive (surpluses in Lk relative to Lk_0, $\Delta Lk > 0$) supercoiling manifests itself in the formation of branched, interwound DNA superhelices (Fig. 2.5). A detailed analysis of the properties of supercoiled DNA based on Monte Carlo simulations of closed wormlike chains has been presented previously [21]. This model depends on only three parameters: P,

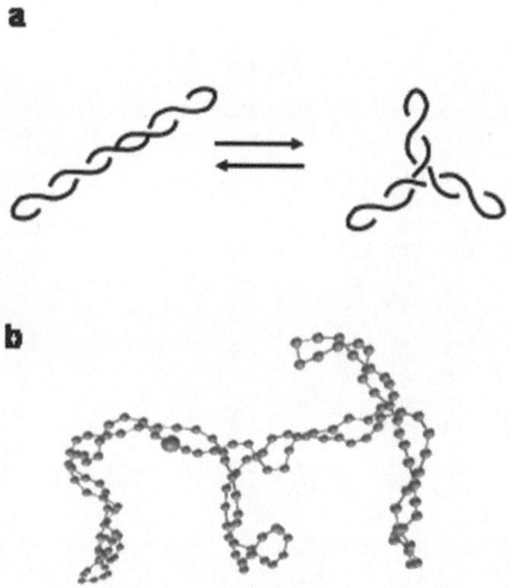

Fig. 2.5. Interconversion of idealized branched and unbranched plectonemic superhelices. (a) An unbranched plectonemic superhelix is in dynamic equilibrium with branched forms, a process that is largely entropy driven. (b) Conformation of a negatively supercoiled 4,600 base-pair plasmid simulated by the algorithm described in [21]. The structure of the plasmid is clearly plectonemic and branched; the ability of the algorithm to reproduce this property accounts for its strong predictive value in computing equilibrium properties of random-sequence plasmids

the torsional rigidity, and an effective excluded-volume diameter for the double helix. The Monte Carlo simulation has been extremely successful in accounting for the bulk of available experimental data on superhelical DNA.

In addition to supercoiling, knotting, and catenation are other biologically important topological states of circular DNA. DNA molecules can become knotted or catenated through the action of topoisomerases and recombinases (reviewed in [22]) and catenated DNAs are obligate intermediates in the replication of circular genomes. A major question in DNA enzymology is therefore how cellular systems acting at the local DNA level sense the topological state of DNA molecules, a global property, and use this information to resolve unfavorable entanglements. Both knotting and catenation of a genome are serious obstacles to normal biological function and fatal to the cell. Hence, all self-knotting and linkage between individual genomes must be completely eliminated. Even a distribution of topological states centered about the unlinked state cannot be tolerated if a species is to be successfully propagated.

Knots and catenanes are characterized by their number and arrangement of minimal or irreducible crossings. For knots, the number of irreducible crossings is denoted Kn and for catenanes the corresponding quantity is Ca. In a

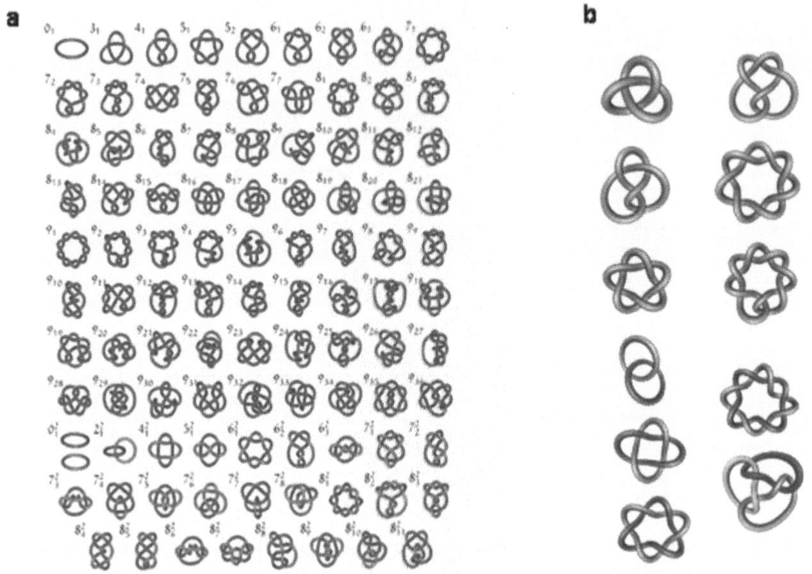

Fig. 2.6. (**a**) The set of all prime topologies containing up to nine irreducible knot crossings and dimeric catenanes with up to eight irreducible crossings generated using the program *Knot Plot*. Each structure is labeled above and to the left according to Alexander and Briggs notation in the case of knots and Rolfsen notation in the case of catenanes. (**b**) Some knots and catenanes of biological importance, left side (from top to bottom): +3 trefoil knot, 4-noded knot, +5-noded torus knot, 2-catenane (or Hopf link), 4-noded torus catenane, 6-noded torus catenane. Right side: 5-noded twist knot, +7-noded torus knot, 7-noded twist knot, 8-noded torus catenane, trimeric (three-component) singly linked catenane

computational analysis, the more than one million possible knots containing up to 16 irreducible crossings have been cataloged [23]. A gallery of all knots with up to nine irreducible crossings and all dimeric catenanes with up to eight crossings is shown in Fig. 2.6 along with several biologically important examples. These species are readily separated by gel electrophoresis (Fig. 2.7).

Only a limited subset of DNA knots and catenanes have been encountered in biological contexts, implying that these topological forms do not result from random linking and unlinking of DNA. Instead, knots and catenanes are generated via pathways that reflect the specific mechanisms of proteins involved in DNA metabolism. The power of DNA topology therefore resides in its ability to illuminate the mechanism of a particular biological process via analysis of the topological forms that these processes generate. The high topological specificity of DNA knotting and linking greatly limits the number of possible mechanistic scenarios and very effectively eliminates implausible ones.

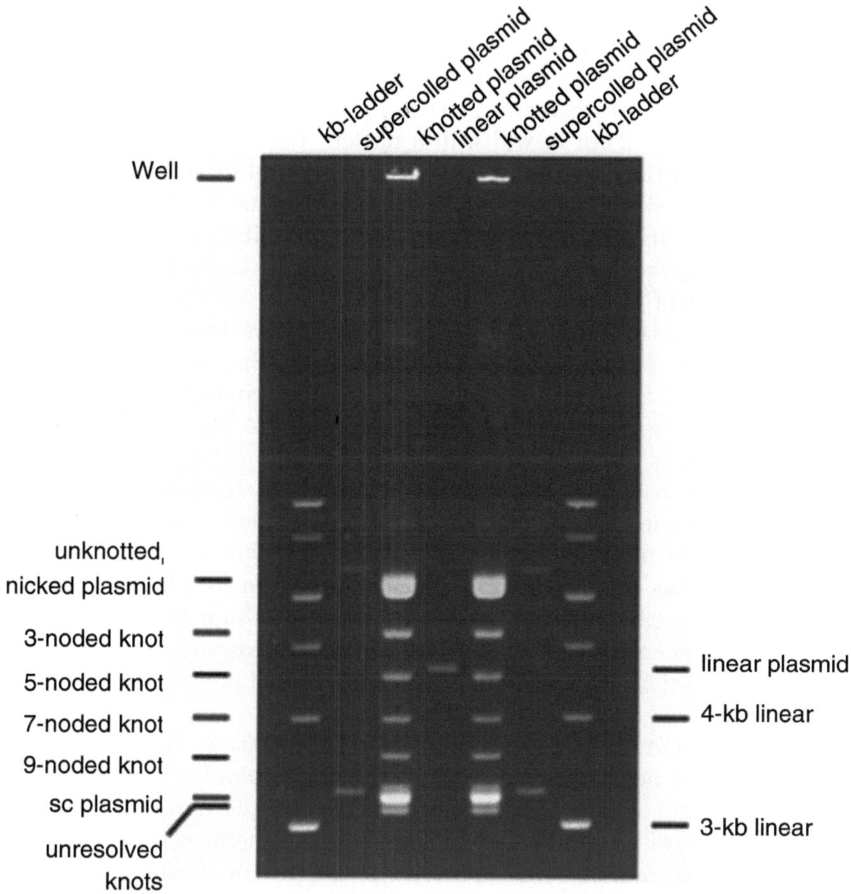

Fig. 2.7. Knotted products generated by phage λ integrative site-specific recombination. A 4,600-bp plasmid bearing λ-att recombination sites was incubated with λ integrase in the presence of IHF protein, nicked with DNase I to remove residual supercoiling, and subjected to electrophoresis on a 1% agarose gel in TBE buffer. Products were made visible by staining with ethidium bromide and the gel image captured using a Peltier-cooled CCD camera. Knotted products appear as bands that are separated by intervals of two nodes; this is consistent with the exclusive formation of knots that belong to the torus family

2.1.5 Flexibility and Topology of DNA, and Their Relation to Genome Organization

In the genomes of prokaryotic and eukaryotic organisms DNA is present in condensed nucleoprotein complexes rather than naked, extended molecules. Moreover, in the nucleii of eukaryotic cells genomes are partitioned among multiple, distinct chromosomes. These aspects of genome organization facilitate the 10^6-fold compaction required to store an enormous amount of genetic information in a cell nucleus that is of order 5–10 μm in diameter. In the case of the human genome, there is about 2 m of DNA per nucleus if the single molecules that comprise each of the 46 chromosomes in a diploid human cell are stretched to their length and placed end to end. Storing a molecule that is 2 m in length and 2 nm in width in a nucleus that is 10 μm in diameter is comparable to storing a string about 60 km long and of cross section 50 μm in an object the size of a basketball. This enormous level of compaction, which is achieved in such a way as to preserve accessibility of the genome for transcription, replication, recombination, and repair, is one of the supremely remarkable feats of biology.

The basic structural unit of organization in eukaryotic chromosomes is the nucleosomal core particle. This is a complex consisting of 147 base pairs of DNA wrapped 1.7 times around a protein core that contains two copies each of the core histone proteins H2A, H2B, H3, and H4 [24, 25]. Nucleosomes are strung together along the DNA, like beads on a string, separated by intervals of 10–80 base pairs of unwrapped DNA. In a chromosome, thousands of these nucleosomes are arranged in a continuous helical array to generate a fiber that is 30 nm in cross section; this 30-nm fiber is in turn folded to generate the higher-level structures that comprise the so-called *chromonema fiber*. Interactions that mediate the association of 30-nm fibers are thought to involve solvent-exposed domains of the core nucleosomes as well as other histone proteins that are associated with nucleosomal arrays, such as H1 and H5 [26, 27].

One critical, but often overlooked, factor that facilitates this compaction is the role played by negative DNA supercoiling. Interwound superhelices, which exemplify the form of superhelical winding that takes place free in solution, are in equilibrium with toroidally wound superhelices generated during wrapping of DNA on the surface of histone proteins in the case of eukaryotic chromatin (Fig. 2.8). In prokaryotic cells, histone-like architectural proteins play similar roles in genome organization [28]. The particular biological advantage achieved by such high levels of organization must have been accompanied by the parallel evolution of mechanisms needed to reorganize local regions of chromatin as required by the cell.

The mechanisms involved in chromatin reorganization have begun to reveal themselves at the molecular level, although many details remain to be worked out. Known as chromatin remodeling, the repositioning of histone octamers on DNA involves interactions with complexes of specific proteins that facilitate the transfer of histones to other available binding sites along the same DNA molecule (in cis) [29]. All known chromatin-remodeling activities appear to require ATP as a source of free energy. These protein complexes often work in concert with enzymes that carry out covalent modifications of histone proteins such as histone transacetylase. This suggests that at least these two activities, and possibly others, are coordinated to make naked DNA available to other cellular factors. There is evidence that acetylation of histones destabilizes histone–histone interactions among individual 30-nm fibers;

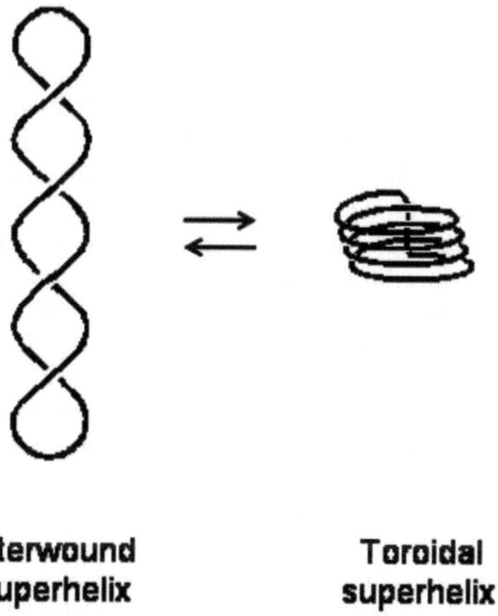

Fig. 2.8. Equilibrium between interwound (plectonemic) and toroidal superhelical structures. The free energy difference between the interwound structure, that found in free solution, and toroidal forms, such as that present in nucleosomes, is modest, which suggests that the structure of negatively supercoiled DNA is dynamic. This equilibrium could easily be perturbed by local motions of nucleosomes on a negatively supercoiled template

these interactions may be responsible for the higher levels of organization in the chromonema fiber [30, 31]. The modifications seem to have negligible effects on the stability of the nucleosome, or on the 30-nm fiber alone. In contrast, chromatin-remodeling complexes appear to act on a local level by forming intermediate complexes with core histones, dissociating the histones from core-particle DNA, and transferring these complexes to available binding sites on naked DNA. The picture that therefore emerges, albeit an oversimplified one, is one in which different cellular factors operate reversibly on specific levels of chromatin organization.

2.1.6 DNA Topology and Enzymology: Flp Site-Specific Recombination

Site-specific recombination is a process that is functionally (but not mechanistically) equivalent to a combination of restriction endonuclease and DNA ligase activities. The recombinases that mediate these recombination events recognize specific DNA sequences and the recombinase-bound sites interact in a nucleoprotein intermediate called *the synaptic complex*. The synaptic complex is responsible for carrying out specific cleavage, strand exchange, and ligation steps that result in the inversion, deletion, or fusion of DNA segments. This biologically essential mode of DNA recombination is involved in gene amplification and copy-number control [32], viral and phage host specificity [33], the generation of antibodies [34, 35], and the transposition of drug-resistance genes [36, 37]. There are also many emerging DNA-delivery applications of site-specific recombination systems in biotechnology such as therapeutic gene targeting, generation of chromosomal translocations, large deletions, and tissue-specific or conditional knockouts as well as site-specific integration, and the precise removal of selectable markers [38, 39].

Flp is a eukaryotic site-specific recombinase from budding yeast (*Saccharomyces cerevisiae*), which is believed to play a role in maintaining the 2-μm circle, a yeast plasmid, at high levels independent of chromosomal copy-number control. This enzyme is responsible for a reaction that causes inversion of a DNA segment near the origin of replication on the 2-μm circle to generate a quasirolling-circle replication intermediate. This intermediate form can lead to the production of many tandem copies of the 2-μm genome, which are subsequently split into individual monomeric circles via a deletion reaction also carried out by the recombinase. Much of our understanding of the mechanism of Flp recombination comes from in vitro studies employing naked DNA and the purified protein. However, as a eukaryotic recombinase, Flp must contend with the presence of nucleosomes and higher order chromatin structure in vivo. Although as yet incompletely understood, some efforts to characterize the in vivo behavior of this system have been reported (see below).

The Flp system is a member of the Int superfamily of site-specific recombinases, so called because of their mechanistic similarities to the integrative recombination system of bacteriophage λ. The phage λ system, which consists of several proteins, is responsible for integration of the bacteriophage genome at a specific location in the *E. coli* chromosome and its subsequent excision at the onset of the lytic stage of the phage life cycle. Some common features of the recombinases in the Int superfamily are the formation of torus knots and catenanes and mechanisms that proceed via an obligate four-stranded DNA structure during recombination, termed *a Holliday junction* (Fig. 2.9). Holliday junctions are key intermediates in a number of other DNA-recombination events, including homologous or general recombination, which, unlike site-specific recombination, can occur at essentially arbitrary locations throughout a genome.

Torus knots and catenanes are so called because these particular forms can be drawn on the surface of a torus. The fact that only products of this topology are formed suggests that the juxtaposition of recombination sites takes place through a "random-collision" mechanism that traps a variable number of superhelical turns between the sites (Fig. 2.10). However, with relaxed circular DNA a bias in the distribution of recombination-product topology indicates that asymmetry exists within the synaptic complex, a feature that is likely to be related to the structure of the Holliday-junction intermediate [40].

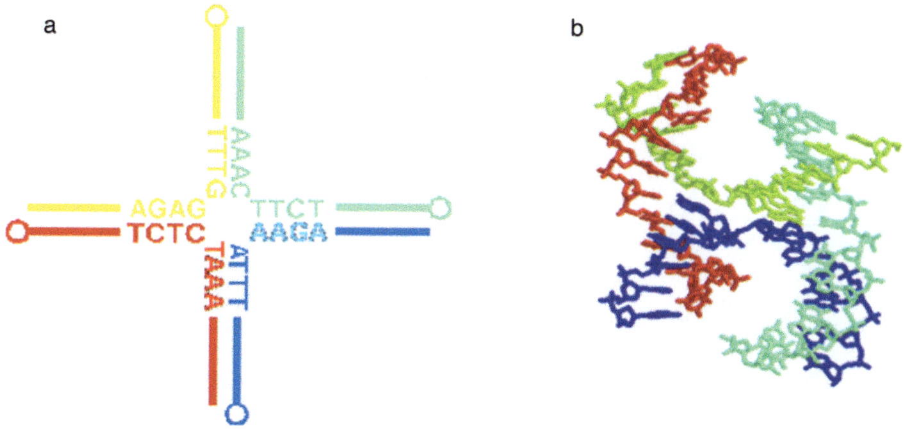

Fig. 2.9. Two representations of the structure of a DNA Holliday junction. (**a**) Diagram showing the alignment of DNA sequences within an immobile Holliday-junction analog. The absence of symmetry abolishes the ability of this junction to undergo branch migration, which would permit relocation of the branch point along the DNA sequence. (**b**) Composite three-dimensional structure of the immobile junction shown in (**a**) based on detailed biophysical studies from several laboratories. The junction assumes a roughly fourfold symmetric structure in which the non-exchanging DNA strands are oriented in an antiparallel fashion

Because of its importance as a recombination intermediate, substantial effort has been focused on elucidating the structure of Holliday junctions. Holliday junctions that are generated by recombination normally have at least twofold DNA-sequence symmetry and thus can undergo isomerization via a process called *branch migration* [41]. Most available data on the structure of four-way DNA junctions has been derived from studies of immobile junctions, which lack the symmetry required for branch migration. Nevertheless, the picture that has emerged from these efforts has been extremely informative and provided a framework for addressing the more complex problem of Holliday-junction intermediates bound to recombination and other junction-recognizing proteins.

One question that we have investigated in our laboratory is that of the geometry of duplex DNA segments in the Flp synaptic intermediate [42]. In particular, we have sought to determine the average relative alignment of duplex recombination sites in the intermediate Flp-DNA complex because biases in relative orientation influence the overall topology of circular Flp-recombination products and the interpretation of topology in terms of recombination mechanism. This is a question that is best addressed by directly imaging intermediate complexes, which we have done by using transmission electron microscopy.

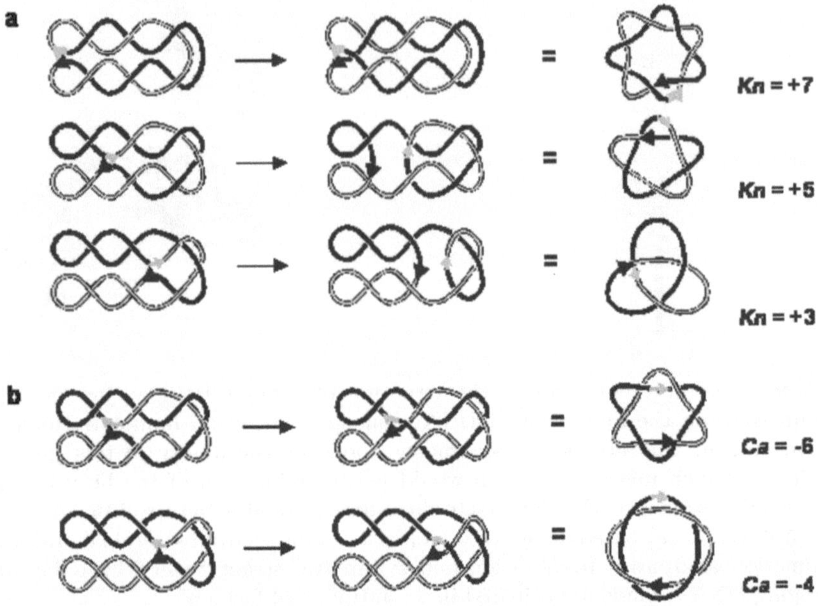

Fig. 2.10. Topology of products generated by recombination of circular DNA molecules mediated by the λ-int superfamily of site-specific recombinases. Diagrams show planar projections of negatively supercoiled DNA substrates undergoing intramolecular recombination. Recombination sites, indicated by *arrows*, divide the DNA contour into two domains, shown as black and outlined gray curves. Random Brownian motion of recombination sites (left column) leads to site synapsis in DNA conformations that involve varying numbers of interdomainal supercoils (supercoils involving separate DNA domains). Only interdomainal supercoils are trapped in the form of knot or catenane crossings by strand-exchange steps in recombination (middle column). The resulting topologies are shown in the form of diagrams (right column) that depict only the number and topological sign of irreducible crossings in each knotted or catenated product, which are given to the right below each figure. These diagrams correspond to actual products in which extraneous supercoils have been removed by nicking of one DNA strand. (**a**) Inversely oriented sites. Inversion generates knotted products that are separated by intervals of +2 knot crossings; these knots belong to the so-called "torus" class because of the property that these knots can all be inscribed on the surface of a torus. Only three examples of knotted products are shown. (**b**) Directly oriented sites. In addition to unlinked circles, deletion reactions generate (−) torus catenanes that also differ by steps of two crossings. Only two examples of catenated products are shown

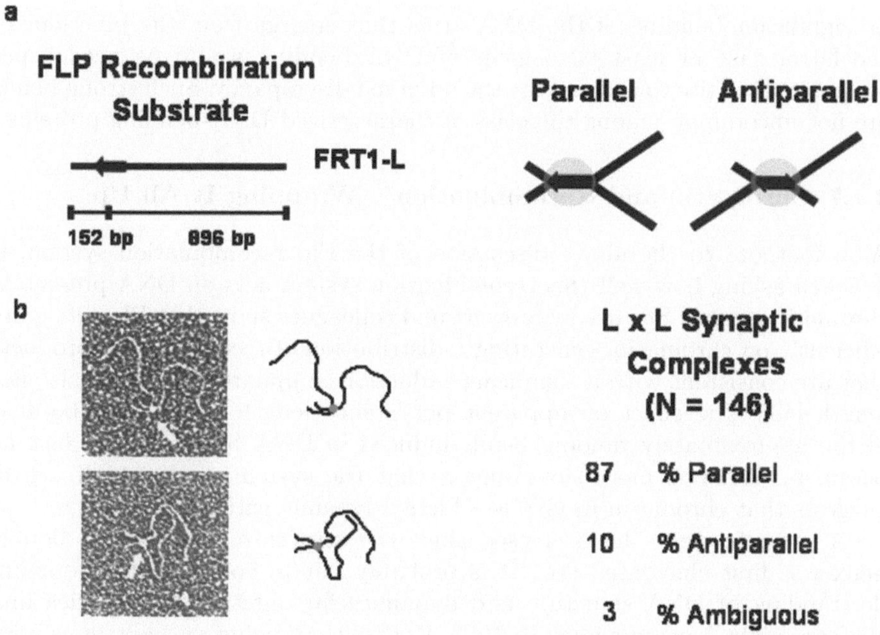

Fig. 2.11. (**a**) Linear DNA recombination substrates consisting of minimal FRT sites present in a specific orientation and in the indicated location. In the presence of Flp these molecules can synapse to form either parallel or antiparallel intermediate complexes. (**b**) Synaptic complexes formed on 1048 bp FRT1-L fragments visualized by electron microscopy and corresponding digital tracings of DNA contours in the micrographs. The *arrow* indicates the position of the FLP synaptosome. Both complexes were scored as parallel. The distribution of parallel and antiparallel complexes is also shown, based on an analysis of 146 synaptic complexes. For details, see [42]

There are two distinct scenarios for DNA duplex juxtaposition in the Flp synaptic intermediate. As shown in Fig. 2.11, synapsis of two DNA segments in which the target site is located close to one end of the molecule could align so that identical pairs of short and long DNA segments that flank the target site are present on either side of the complex. This structure would imply that the global alignment of target sites is roughly parallel. Alternatively, pairing of segments that are dissimilar in size implies the formation of a globally antiparallel synapse. Analysis of over 100 synaptic complexes shows a strong preference for globally parallel alignment of the recombination sites, as shown in the examples in Fig. 2.10.

The observed preference for globally parallel alignment of recombination sites stands in contrast to the expected antiparallel alignment of sites based on the geometry of immobile Holliday-junction analogs. It is possible that this difference is due to dramatic differences in the geometry of the four-way junction intermediate in the bound-Flp state; alternatively, there may be significant bending of the DNA arms that emerge from the junction. In the latter case at least 90 degrees of DNA bending would be needed per recombination site to account for the observed discrepancy. Such strong bends are not uncommon among the class of characterized DNA-bending proteins.

2.1.7 Chromatin and Recombination – Wrapping It All Up

As a footnote to the above discussion of the Flp recombination system, it is worth asking how well this recombination system acts on DNA present as chromatin in vivo. Studies by Stewart and colleagues show that Flp acts quite efficiently on chromatin, generating a distribution of recombination products that are consistent with a significant reduction in apparent DNA persistence length [43]. The effect on apparent persistence length is expected because of the approximately random bends induced in DNA by binding of histone octamers. What is more surprising is that the system works at all, which suggests that chromatin in vivo is a highly dynamic entity.

A mere 50 years have elapsed since the structure of the DNA double helix was first elucidated [44]. It is probably fair to speculate that our understanding of DNA structure and dynamics far outstrips what Crick and Watson might have imagined in 1953. Rather than being an inert repository of genetic information, the DNA molecule is capable of taking on a wide range of sequence-dependent structures, some of which exert profound effects on the molecule's accessibility and its participation in activities such as transcription, replication, recombination, and repair. The genomic DNA of eukaryotes is a particularly dynamic entity in vivo whose structure is subject to continual modification through the displacement of nucleosomes and interactions with chromatin-remodeling complexes. Since the 30,000 or so genes in the human genome account for only a small fraction of the genome's information content, it is likely that there is much that remains to be learned about the structure and function of the bulk of human DNA.

References

1. V.A. Bloomfield, D.M. Crothers, I.J. Tinoco, *Nucleic Acids: Structures, Properties and Functions* (University Science Books, Herndon, VA, 2000)
2. A. Rich, Ann., N Y Acad. Sci. **726**, 1–16; discussion 16–17 (1994)
3. A. Rich, S. Zhang, Nat. Rev. Genet. **4**, 566–572 (2003)
4. E. Selsing, R.D. Wells, C.J. Alden, S. Arnott, J. Biol. Chem. **254**, 5417–5422 (1979)
5. J.G. Nadeau, D.M. Crothers, Proc. Natl. Acad. Sci. USA **86**, 2622–2626 (1989)
6. D.M. Crothers, T.E. Haran, J.G. Nadeau, J. Biol. Chem. **265**, 7093–7096 (1990)
7. O. Kratky, G. Porod, Rec. Trav. Chim. Pays-Bas **68**, 1106–1122 (1949)
8. R.T. Kovacic, K.E. van Holde, Biochemistry **16**, 1490–1498 (1977)
9. P.J. Hagerman, Biopolymers **20**, 1503–1535 (1981)
10. W.H. Taylor, P.J. Hagerman, J. Mol. Biol. **212**, 363–376 (1990)

11. D.M. Crothers, J. Drak, J.D. Kahn, S.D. Levene, Methods Enzymol. **212**, 3–29 (1992)
12. D. Lang, H. Bujard, B. Wolff, D. Russell, J. Mol. Biol. **23**, 163–181 (1967)
13. C. Rivetti, C. Walker, C. Bustamante, J. Mol. Biol. **280**, 41–59 (1998)
14. S.D. Levene, *Nature Encyclopedia of Life Sciences* (2001)
15. J.A. Schellman, S.C. Harvey, Biophys. Chem. **55**, 95–114 (1995)
16. J.C. Venter, et al. Science **291**, 1304–1351 (2001)
17. G. Calugareanu, Rev. Math. Pures. Appl. **4**, 5–20 (1959)
18. J.H. White, Am. J. Math. **91**, 693–728 (1969)
19. F.M. Pohl, J. Math. Mech. **17**, 975–985 (1968)
20. J.H. White, W.R. Bauer, J. Mol. Biol. **189**, 329–341 (1986)
21. A.V. Vologodskii, S.D. Levene, K.V. Klenin, M. Frank-Kamenetskii, N.R. Cozzarelli, J. Mol. Biol. **227**, 1224–1243 (1992)
22. N.R. Cozzarelli, M.A. Krasnow, S.P. Gerrard, J.H. White, Cold Spring Harb. Symp. Quant. Biol. **49**, 383–400 (1984)
23. J. Hoste, M. Thistlethwaite, J. Weeks, Math. Intelligencer **20**, 33–48 (1998)
24. K. Luger, A.W. Mader, R.K. Richmond, D.F. Sargent, T.J. Richmond, Nature **389**, 251–260 (1997)
25. K. Luger, T.J. Richmond, Curr. Opin. Struct. Biol. **8**, 33–40 (1998)
26. J. Widom, Curr. Biol. **8**, R788–791 (1998)
27. J.J. Hayes, J.C. Hansen, Curr. Opin. Genet. Dev. **11**, 124–129 (2001)
28. C.J. Dorman, P. Deighan, Curr. Opin. Genet. Dev. **13**, 179–184 (2003)
29. C.J. Fry, C.L. Peterson, Curr. Biol. **11**, R185–197 (2001)
30. A.T. Annunziato, J.C. Hansen, Gene Expr. **9**, 37–61 (2000)
31. P.J. Horn, C.L. Peterson, Science **297**, 1824–1827 (2000)
32. F.C. Volkert, J.R. Broach, Cell **46**, 541–550 (1986)
33. P. van de Putte, N. Goosen, Trends Genet. **8**, 457–462 (1992)
34. D.B. Roth, T. Lindahl, M. Gellert, Curr. Biol. **5**, 496–499 (1995)
35. A. Agrawal, Q.M. Eastman, D.G. Schatz, Nature **394**, 744–751 (1998)
36. J.R. Scott, G.G. Churchward, Annu. Rev. Microbiol. **49**, 367–397 (1995)
37. C.A. Liebert, R.M. Hall, A.O. Summers, Microbiol. Mol. Biol. Rev. **63**, 507–522 (1999)
38. M.M. Golic, Y.S. Rong, R.B. Petersen, S.L. Lindquist, K.G. Golic, Nucleic Acids Res. **25**, 3665–3671 (1997)
39. B.D. Bethke, B. Sauer, Methods Mol. Biol. **133**, 75–84 (2000)
40. N.J. Crisona, R.L. Weinberg, B.J. Peter, D.W. Sumners, N.R. Cozzarelli, J. Mol. Biol. **289**, 747–775 (1999)
41. I.G. Panyutin, P. Hsieh, Proc. Natl. Acad. Sci. USA **91**, 2021–2025 (1994)
42. K.E. Huffman, S.D. Levene, J. Mol. Biol. **286**, 1–13 (1999)
43. L. Ringrose, S. Chabanis, P.O. Angrand, C. Woodroofe, A.F. Stewart, EMBO. J. **18**, 6630–6641 (1999)
44. J.D. Watson, F.H. Crick, Nature **171**, 737–738 (1953)
45. S.D. Levene, H.M. Wu, D.M. Crothers, Biochemistry **25**, 3988–3995 (1986)
46. H.S. Koo, J. Drak, J.A. Rice, D.M. Crothers, Biochemistry **29**, 4227–4234 (1990)

3

Monte Carlo Simulation of DNA Topological Properties

A. Vologodskii

Summary. In the scale of hundreds and thousands of base pairs, DNA double helix is a very flexible polymer chain that adopts many different conformations in solution. The properties of such molecules have to be analyzed in terms of statistical mechanics. Now these properties can be simulated with very good accuracy. Here we review this simulation technique, with emphasis on topological properties of circular DNA. We describe the basic concepts related with DNA topological properties and illustrate, by comparing simulation results with the experimental data, how accurately these properties can be computed. We consider DNA model used in the simulation, methods of sampling of the statistical ensemble, simulation of DNA supercoiling, and different problems, related with knots and links in circular DNA. To analyze topological state of closed chain one needs to calculate a topological invariant. We describe the algorithms that allow one to compute one of such invariants, Alexander's polynomial, which is especially suitable for the Monte Carlo simulation. At the end, we consider special methods of sampling for rare DNA conformations.

3.1 Introduction

It became clear in the last few years that large-scale conformational properties of DNA can be simulated with very good accuracy. These simulations reproduce experimental data on hydrodynamic properties of DNA molecules [1–3], DNA cyclization [4–7], equilibrium distributions of topological states [8–13], elasticity of the single molecules [14, 15] and light and neutron scattering data on supercoiled DNA [16–19]. The simulations are based on the statistical–mechanical treatment of a well-established model of the double helix. All parameters of the model have been reliably determined for various solution conditions. Thus, the simulations are capable of providing reliable quantitative information on many DNA properties that are hardly measurable experimentally. They became an important instrument in the studies of different protein systems that interact simultaneously with two or more DNA sites. On the other hand, DNA molecules represent an ideal object for polymer physics, and especially for studying topological properties of polymer chains. They are homogeneous, thin, and long. They can be easy converted from linear to circular form. Circular DNA molecules in different topological states can be separated by gel electrophoresis and, thus, the distributions of topological states can be studied experimentally. The methods of experimental manipulation with DNA molecules, developed in the last few decades, allows one to easily perform many things that are beyond the imagination of traditional polymer chemistry and physics. These features of DNA molecules can be used to study general properties of polymer chains, and our ability to simulate these properties with high accuracy helps greatly in such studies. The computational methods, which allow one to simulate large-scale statistical properties of DNA, are a subject of this review. Major attention is paid to the simulations to topological properties of circular DNA. Therefore we begin the review with a brief description of the basic concepts related to DNA topology. Then we illustrate, by comparison with the experimental data, how well conformational properties of circular DNA can be computed. Detailed description of the computational methods will follow the analysis of the basic DNA model. The review is restricted by Monte Carlo simulation of the equilibrium DNA properties, which has a wider use than the dynamic simulation based on the Brownian dynamics method [20–23], for example).

3.2 Circular DNA and Supercoiling

The circular form of DNA is widespread in nature. In this form each of the two strands that make up the DNA molecule is closed in on itself. A diagrammatic view of closed circular DNA is presented in Fig. 3.1.

The two strands of the double helix in closed circular DNA are linked. In topological terms, the links between the strands of the double helix belong to

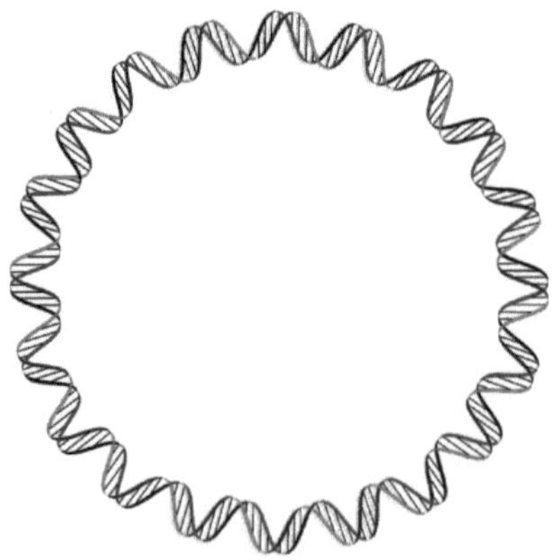

Fig. 3.1. Diagram of closed circular DNA. Two strands of the double helix are shown together with the base pairs which are perpendicular to the helix axis. The linking number of the complementary strands, Lk, equals 20

the torus class. The quantitative description of such links is called *the linking number* (Lk), which may be determined in the following way. One of the strands defines the edge of an imaginary surface (any such surface gives the same result). The Lk is the algebraic (i.e., sign-dependent) number of intersections between the other strand and this spanning surface. By convention, the Lk of a closed circular DNA formed by a right-handed double helix is positive. Lk depends only on the topological state of the strands and hence is maintained through all conformational changes that occur in the absence of strand breakage. Its value is always integral. Lk can be also defined through the Gauss integral:

$$\mathrm{Lk} = \oint_{C_1}\oint_{C_2} \frac{(d\mathbf{r}_1 \times d\mathbf{r}_2)\mathbf{r}_{12}}{r_{12}^3}, \tag{3.1}$$

where $\mathbf{r}_1$ and $\mathbf{r}_2$ are vectors whose ends run, upon integration, over the first and second contours, C_1 and C_2, respectively, $\mathbf{r}_{12} = \mathbf{r}_2 - \mathbf{r}_1$.

Quantitatively, the linking number of the complementary strands is close to N/γ, where N is the number of base pairs in the molecule and γ is the number of base pairs per double-helix turn in linear DNA under given conditions. However, these values are not exactly equal one to another. The difference between $\mathbf{r}_{12} = \mathbf{r}_2 - \mathbf{r}_1$ and N/γ, *the linking number difference*, $\Delta\mathrm{Lk}$, defines most of the properties of closed circular DNA:

$$\Delta\mathrm{Lk} = \mathrm{Lk} - N/\gamma. \tag{3.2}$$

The value of $\Delta\mathrm{Lk}$ is not a topological invariant. It depends on the solution conditions that determine γ. Even though γ itself changes very slightly with changing ambient conditions, these changes may substantially alter $\Delta\mathrm{Lk}$, as the right-hand part of (3.2) is the difference between two large quantities that are close in value.

It often proves more convenient to use the value of superhelix density, $\Delta\mathrm{Lk}$, which is $\Delta\mathrm{Lk}$ normalized for the average number of helical turns in nicked circular DNA, N/γ:

$$\sigma = \Delta\mathrm{Lk} \cdot \gamma/N. \tag{3.3}$$

Whenever $\Delta\mathrm{Lk} \neq 0$, closed circular DNA is said to be supercoiled. The entire double helix is stressed in this case. This stress can either lead to a change in the actual number of base pairs per helix turn in closed circular DNA or cause regular spatial deformation of the helix axis. The axis of the double helix then forms a helix of a higher order, superhelix (Fig. 3.2).

It is this deformation of the helix axis in closed circular DNA that gave rise to the term *superhelicity* or *supercoiling* [24]. Circular DNA extracted from cells turns out to be always (or nearly always) negatively supercoiled and has a $\Delta\mathrm{Lk} \neq 0$ between -0.03 and -0.09, but typically near the middle of this range [25].

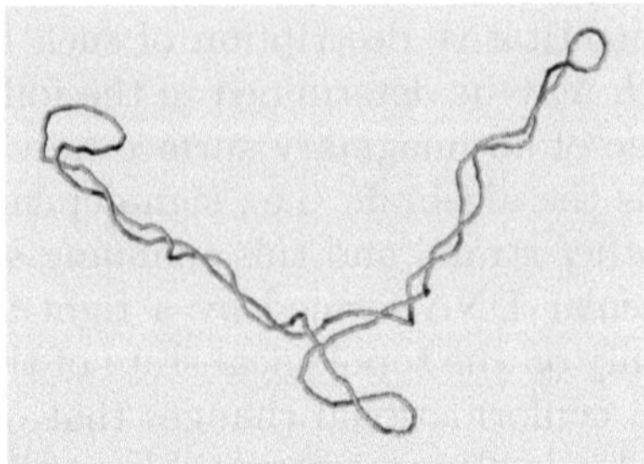

Fig. 3.2. A typical conformation of supercoiled DNA. The *double helix* is presented here by the flexible rod. The picture obtained by computer simulation of supercoiled molecule 3,500 base pairs in length, $\Delta \text{Lk} \neq 0 = -0.06$, for physiological ionic conditions

Supercoiling can be structurally realized in two ways: by deforming the molecular axis and by altering the twist of the double helix. Quantitatively it can be expressed by White's theorem:

$$\Delta \text{Lk} = \Delta \text{Tw} + \text{Wr}, \tag{3.4}$$

where ΔTw is the difference between actual twist of DNA and the average value of twist in nicked form of the same DNA, and Wr is writhe of the double helix. The value of Wr is defined by the spatial course of DNA axis only, it is the property of simple closed curve. Wr can be thought as a measure of a curve's net right-handed or left-handed asymmetry, i.e., its chirality, and is therefore zero for a planar curve. The value of Wr is equal to the Gauss integral (3.1), in which integration is performed both times along the same contour – the DNA axis. Thus, Wr can be represented as a sum of two values that corresponds to the available degrees of freedom: the torsional deformation of the double helix and deformation of DNA axis. Detailed description of Wr properties can be found in references [26–29].

If circular DNA has a single-stranded nick, any torsional stress of the double helix disappears quickly. Although such DNA molecules cannot be supercoiled, they can form different knot and links. The Topology of the nicked molecule is completely specified by conformation of its axis.

3.3 Testing the DNA Model

In the middle of 1990s we knew a lot about large-scale conformational properties of DNA. There were convincing data that indicated that the equilibrium conformations of linear and nicked circular DNA could be described quantitatively in terms of the wormlike chain model, that also accounts for the electrostatic interaction between DNA segments [9, 10, 13]. However, it was not clear until that time how well the model could describe the conformational

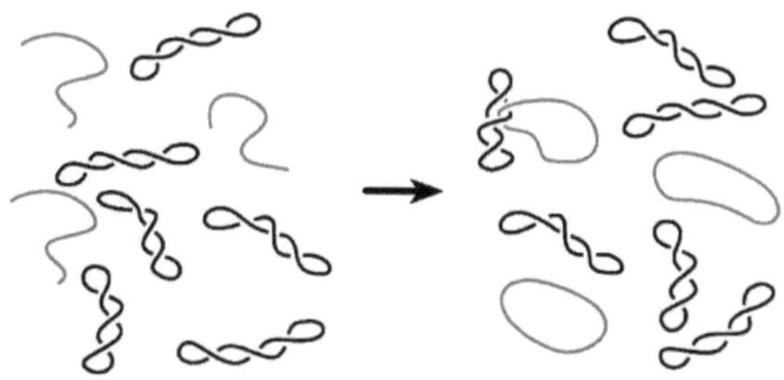

Fig. 3.3. Probing conformational properties of supercoiled DNA by formation of topological links [11]. The diagram shows formation of links between supercoiled and cyclizing linear molecules. The cyclization occurred via long cohesive ends and resulted in nicked circular molecules. The simplest links shown here comprised at least 90% of all links formed

properties of supercoiled DNAs, a form that is characterized by frequent close approaches between distal DNA segments. Because these close approaches are rare in linear and open circular DNA molecules, the accurate description of intersegment interaction is not so crucial for the prediction of most of their properties. Thus, it was important to test how well the model can describe conformational properties of supercoiled DNA. We completed this task by comparing computed and measured equilibrium linkage between supercoiled DNA and cyclizing linear molecules as illustrated in Fig. 3.3.

The probability P that a given open circular DNA will be linked with supercoiled molecules of concentration Wr can be expressed as

$$P = \int_0^{\infty} p(R) 4\pi r^2 c \frac{N_{\mathrm{A}}}{M} \, \mathrm{d}r, \tag{3.5}$$

where Wr is the probability of linking of these two molecules if their centers of mass are separated by distance R, N_{A} is Avogadro's number, and M is the molecular weight of DNA. The equation has a simple interpretation because the term Wr is the probability of finding a supercoiled molecule in the volume element Wr. Equation (3.5) assumes that the concentration Wr is small enough so that we can ignore the formation of three or more linked molecules. The probability Wr must be averaged over all possible conformations and orientations of the two chains. We used the Monte Carlo method, described in detail below, to calculate Wr. Figure 3.4 shows Wr computed for two circular DNA molecules.

For a comparison with experimental data, it is convenient to introduce the constant B:

$$B = \frac{N_{\mathrm{A}}}{M} \int_0^{\infty} p(R) 4\pi r^2 \, \mathrm{d}r. \tag{3.6}$$

This allows us to express P as

$$P = Bc. \tag{3.7}$$

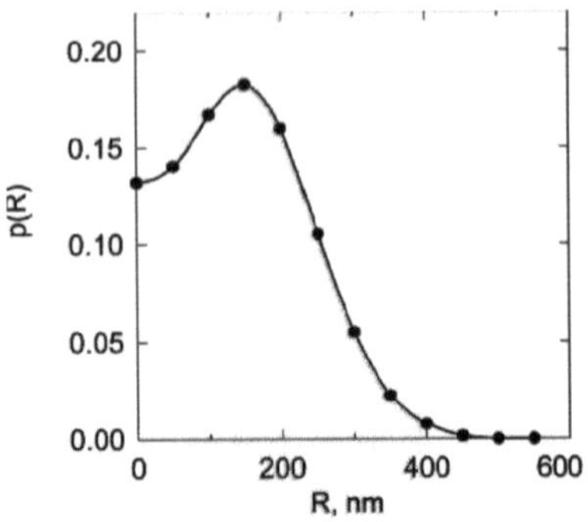

Fig. 3.4. Probability of linkage between two relaxed DNA molecules 7,000 and 10,000 base pairs in length as a function of the distance between the chains centers of mass, R. The function strongly depends on solution ionic conditions. $p(R)$ shown in the figure corresponds to near physiological ionic conditions. For each value of R a fraction of conformations of two chains is forbidden because of overlap of one chain with the other. Since this fraction increases sharply as R diminishes, there is a decrease of $p(R)$ at small values of R

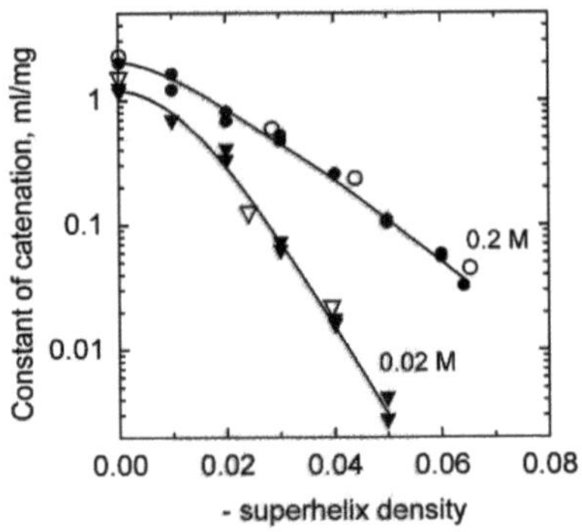

Fig. 3.5. Measured and simulated probabilities of catenation as a function of supercoiling [11]. The experimental values of B (*open symbols*) are shown together with calculated results (*filled symbols*) for NaCl concentrations of 0.02 M (Ü,˙P), 0.2 M (E, J). The large changes of conformational properties of supercoiled DNA with ionic conditions result, in good approximation, from the change of intersegment electrostatic interactions, specified by the variation DNA effective diameter (see below)

The value of B does not depend on DNA concentration and reflects only the properties of a particular circular DNA. It depends on the lengths of both DNAs, the superhelix density, and on the ionic conditions, which change the conformational properties of supercoiled DNA.

Figure 3.5 presents the measured and simulated values of B for solutions of two different concentrations of NaCl (0.02 M, 0.2 M) as a function of the DNA superhelical density [11]. Note that the simulated and measured values of B

agree extremely well over the whole range of σ and NaCl concentrations studied, even though the range of B values exceeds two orders of magnitude. The data make it clear that conformations of supercoiled DNA vary greatly over this range of sodium ion concentrations. This work convincingly proved that the simulation is capable of accurately predicing conformational properties of both relaxed and supercoiled DNA molecules.

3.4 DNA Model

The DNA model represents a discrete analog of the wormlike chain that also accounts for DNA torsional rigidity, excluded volume, and intersegment electrostatic interaction [8, 30–32]. A DNA molecule composed of n Kuhn statistical lengths is modeled as a closed chain consisting of n rigid segments that are cylinders of equal length where k is a computational parameter of our choice (Fig. 3.6).

The bending elastic energy of the chain, n, is computed as

$$E_{\mathrm{b}} = k_{\mathrm{B}} T_{\mathrm{g}} \sum_{i=1}^{kn} \theta_i^2, \tag{3.8}$$

where the summation extends over all the joints between the elementary segments,

$$E_{\mathrm{b}} = \frac{g}{2} \sum_{i=1}^{kn} \theta_i^2$$

is the angular displacement of segment

$$E_{\mathrm{b}} = \frac{g}{2} \sum_{i=1}^{kn} \theta_i^2$$

relative to segment

$$E_{\mathrm{b}} = \frac{g}{2} \sum_{i=1}^{kn} \theta_i^2,$$

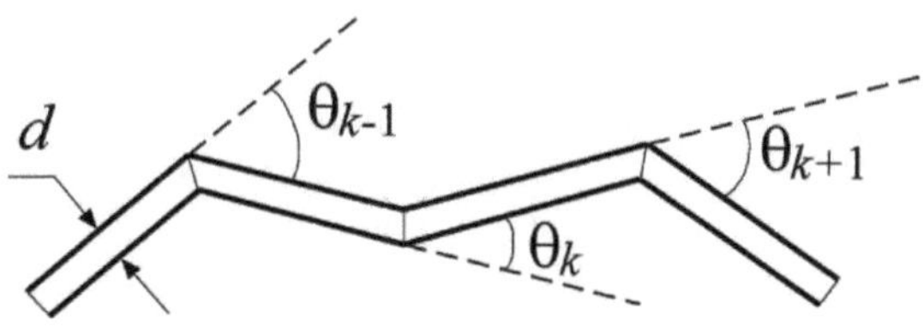

Fig. 3.6. The model of double-stranded DNA. The length of the cylinders can vary, although it usually equals 30 base pairs of the double helix (1/5 of the persistence length)

and

$$E_{\mathrm{b}} = \frac{g}{2}\sum_{i=1}^{kn}\theta_i^2$$

is the bending rigidity constant, $k_{\mathrm{B}}T$ is the Boltzmann temperature factor. The bending constant $k_{\mathrm{B}}T$ is defined so that the Kuhn statistical length corresponds to $k_{\mathrm{B}}T$ rigid segments [30]:

$$k = \frac{1+\langle\cos\theta\rangle}{1-\langle\cos\theta\rangle}, \tag{3.9}$$

where

$$\langle\cos\theta\rangle = \frac{\int_0^\pi \cos\theta\sin\theta\exp(-g\theta^2)\mathrm{d}\theta}{\int_0^\pi \sin\theta\exp(-g\theta^2)\mathrm{d}\theta}. \tag{3.10}$$

The value

$$\langle\cos\theta\rangle = \frac{\int_0^\pi \cos\theta\sin\theta\exp(-g\theta^2)\mathrm{d}\theta}{\int_0^\pi \sin\theta\exp(-g\theta^2)\mathrm{d}\theta}$$

can be found as by numerical solution of (3.10).

Replacement of the continuous wormlike chain with a discrete chain consisting of

$$\langle\cos\theta\rangle = \frac{\int_0^\pi \cos\theta\sin\theta\exp(-g\theta^2)\mathrm{d}\theta}{\int_0^\pi \sin\theta\exp(-g\theta^2)\mathrm{d}\theta}$$

hinged rigid segments is an approximation that improves as

$$\langle\cos\theta\rangle = \frac{\int_0^\pi \cos\theta\sin\theta\exp(-g\theta^2)\mathrm{d}\theta}{\int_0^\pi \sin\theta\exp(-g\theta^2)\mathrm{d}\theta}$$

increases. The computer time needed for a simulation increases approximately as

$$\left(\langle\cos\theta\rangle = \frac{\int_0^\pi \cos\theta\sin\theta\exp(-g\theta^2)\mathrm{d}\theta}{\int_0^\pi \sin\theta\exp(-g\theta^2)\mathrm{d}\theta}\right)^2 .$$

It is therefore necessary to choose a value of

$$\langle\cos\theta\rangle = \frac{\int_0^\pi \cos\theta \sin\theta \exp(-g\theta^2)\mathrm{d}\theta}{\int_0^\pi \sin\theta \exp(-g\theta^2)\mathrm{d}\theta}$$

that is large enough to ensure reliable results but small enough to keep the computational time reasonable. The minimal value of k, which provides the limiting properties of the wormlike chain, depends on a property of interest. Figure 3.7a shows dependence of the average $\langle \mathrm{Wr}\rangle / \Delta\mathrm{Lk}$ for highly supercoiling DNA as a function of k. Clearly, the results for $k \geq 10$ are nearly independent of k. So, $k = 10$ can be used for modeling DNA supercoiling. In this case one straight segment of the model chain corresponds to ≈30 bp (Kuhn statistical length of the double helix corresponds to ≈300 bp). For this value of k the bending rigidity, $\langle \mathrm{Wr}\rangle / \Delta\mathrm{Lk}$, equals 2.403. For another property of circular chains, equilibrium probability of trefoil knots, the results for $k = 1$ and $k = 10$ are hardly distinguishable (Fig. 3.7b). There are some properties, however, which require much larger values of k [14].

The excluded volume effect and electrostatic interactions between DNA segments are taken into account in the model via the concept of effective diameter, $\langle \mathrm{Wr}\rangle / \Delta\mathrm{Lk}$. This is the diameter of impenetrable uncharged cylindrical segments of the model chain. The quantitative definition of $\langle \mathrm{Wr}\rangle / \Delta\mathrm{Lk}$ is based on the concept of the second virial coefficient [33]. It was shown that approximation of the electrostatic interaction by this hard core potential

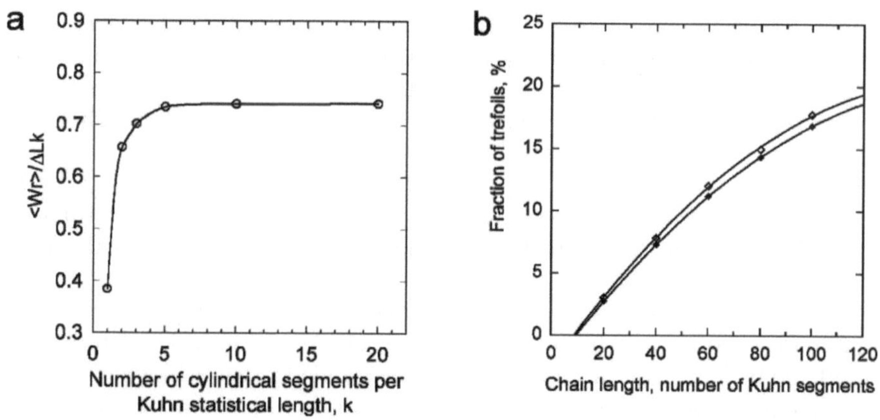

Fig. 3.7. Simulation results as function of k, the number of straight segments per Kuhn length of the model chain. (**a**) Calculated value of $\langle \mathrm{Wr}\rangle / \Delta\mathrm{Lk}$ for highly supercoiled DNA (3,500 base pairs in length, superhelix density of –0.05, effective diameter of the chain equals 5 nm). (**b**) Probability of trefoil calculated for freely jointed chain, $k = 1$, (◇) and $k = 10$ (○)

and by the corresponding Debye–Hückel potential gives very similar results in Monte Carlo simulations of DNA equilibrium properties, with a certain exception for conformations of supercoiled DNA at low concentration of monovalent ions (≤ 0.02 M) [31].

The model's features specified above are sufficient to simulate large-scale DNA conformational properties, both for linear DNA and circular DNA with a single-stranded nick (nicked DNA), because in these cases the conformations of the DNA axis do not depend on the DNA twist, ≤ 0.02 M (if the DNA is intrinsically straight). However, dependence on ≤ 0.02 M is crucial for properties of closed circular DNA. To use the model in this case one can express the displacement of chain twist from its equilibrium value, ≤ 0.02 M, by the equation:

$$\Delta \mathrm{Tw} = \Delta \mathrm{Lk} - \mathrm{Wr}, \tag{3.11}$$

where writhe, Wr, is a property of the chain axis alone [27], and ≤ 0.02 M is the linking number difference of the simulated DNA. The value of ≤ 0.02 M should be considered as a parameter at each simulation [32]. Hence, in this model, the torsional energy, ≤ 0.02 M, is defined by the conformation of the DNA axis and may be expressed as

$$E_\mathrm{t} = (2\pi^2 C/L)(\mathrm{Lk} - \mathrm{Wr})^2, \tag{3.12}$$

where ≤ 0.02 M is the torsional rigidity constant, and L is the DNA length.

There are three parameters of the model that specify equilibrium properties of the double helix; each of these has been determined from independent studies. The first parameter, the Kuhn statistical length (which defines the bending rigidity ≤ 0.02 M), is close to 100 nm for solutions containing more than 0.01 M monovalent ions or more than 1 mM multivalent ions [7, 34]. The second parameter is the DNA torsional rigidity, ≤ 0.02 M. The value of 3×10^{-19} erg cm for ≤ 0.02 M seems to be the most reliable for this range of ionic conditions [9, 35]. The third parameter, the DNA effective diameter, ≤ 0.02 M, depends strongly on ionic conditions. Accurate values of ≤ 0.02 M have been determined in the experimental and theoretical studies (Fig. 3.8) [10, 12, 33, 36, 37].

The model described above is the simplest one that can provide a quantitative description for the large number of DNA conformational properties considered in this proposal. It can be easily extended to cases in which a group of the chain segments forms a specific conformation induced by binding a protein. However a more elaborate model that explicitly accounts for torsional orientation of each segment is required for the analysis of DNA molecules with two or more groups of bent segments distributed along the chain contour. Such a model and the corresponding simulation procedures have been developed [38, 39].

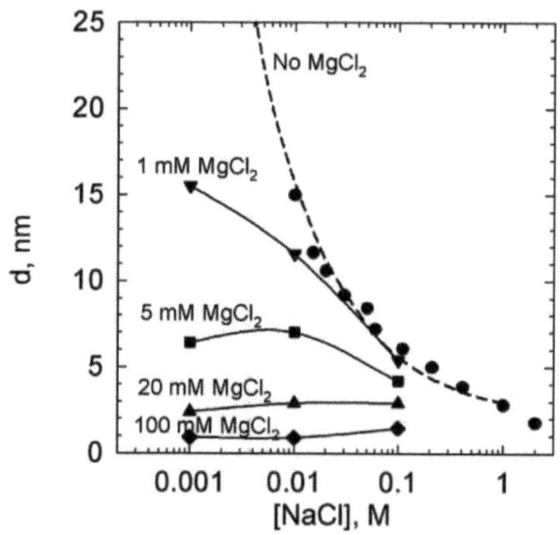

Fig. 3.8. Effective diameter of the model chain as a function of ionic conditions

3.5 Analysis of Topological State for a Particular Conformation

3.5.1 Knots

In many cases we have to determine the topology of particular chain conformations, which can be unknotted or form a particular type of knot, or be linked with other molecules. We need this to keep a topological state of the chain(s) unchanged during a simulation run. Since segments are allowed to pass through each other during the Metropolis procedure (see below), it is necessary to check that the topology of a trial conformation is the same as the current one. Thus, the topology of each trial conformation is calculated and the conformation is rejected if its topology is different from that of the current conformation. We need to determine the topology when we calculate the distribution of topological states, equilibrium or resulting from simulated reactions, catalyzed by enzymes. To determine the topology of a particular conformation of an isolated closed chain, one can calculate the Alexander polynomial, $\leq 0.02\,\mathrm{M}$ [40]. The Alexander polynomial is a topological invariant that describes the knot type of a closed curve (see [41], for example). It has the same value for all topologically equivalent curves, and any two curves with a different value of $\leq 0.02\,\mathrm{M}$ have different topology (Fig. 3.9).

The value $2t^2-3t+2$ for a particular chain conformation can be calculated by the following way ... [40]. First, one projects the knot on a plane along an arbitrarily chosen axis, while drawing breaks at the crossing points in the part of the curve that lies below the other part (Fig. 3.10).

The projection of the knot amounts to the set of curves, which are called *the generators*. Let us arbitrarily choose the direction of passage of the

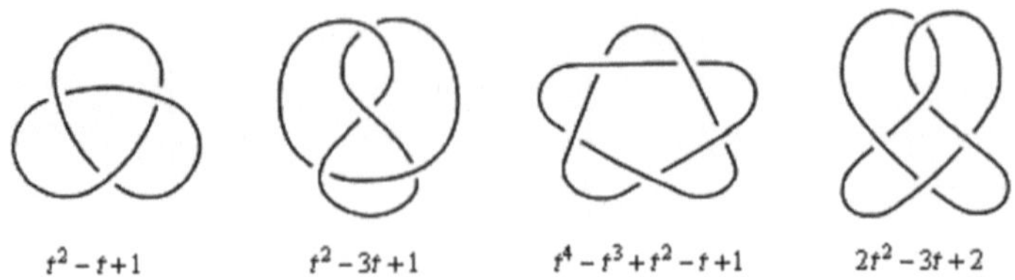

Fig. 3.9. The simplest *knots* and their Alexander polynomials, $\Delta(t)$. All four knots that can be drawn with less than six intersections are shown. For an unknotted contour $\Delta(t) = 1$

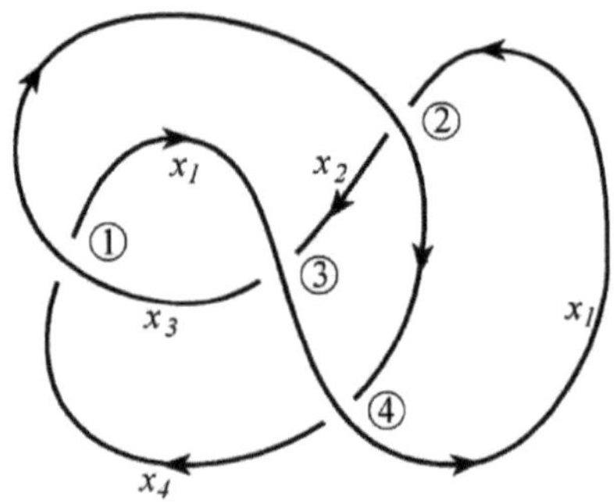

Fig. 3.10. On the calculation of an Alexander polynomial for *knots*. Here x_1, x_2, x_3, and x_4 are the generators, and 1, 2, 3, and 4 are the crossing points in the projection of the knot

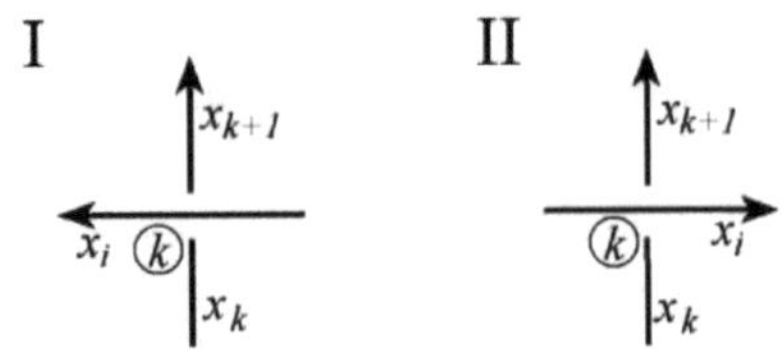

Fig. 3.11. The two types of crossings

generators and renumber them, having selected arbitrarily the first generator. The crossing that separates the kth and $(k+1)$th generators will be called the kth crossing. The crossings are of two types (Fig. 3.11). Thus each crossing is characterized by its number, by its type (I or II), and by the number of generator passing over it.

Now the knot can be correlated with a square Alexander matrix, in which the kth row corresponds to the kth crossing and which consists of N elements (N is the total number of crossings in the projection of the knot). Here all the elements except a_{kk}, a_{kk+1} and a_{ki} (i is the number of the overpassing generators) are zero. The nonzero elements of the kth row are determined as follows:

(1) When $i = k$ or $i = k + 1$, then $a_{k,k} = -1$ and $a_{k,k+1} = 1$, independent of the type of crossings
(2) When $i \neq k$ and $i \neq k + 1$, then
$a_{kk} = 1,\ a_{k,k+1} = -t$ and $a_{kk} = t - 1$ for a type I crossing,
$a_{kk} = -t,\ a_{k,k+1} = 1$ and $a_{kk} = t - 1$ for a type II crossing.

These relationships hold under condition that for $k = N$ one makes the substitution $k + 1 \rightarrow 1$.

To discriminate closed chains into two categories, knotted versus unknotted, it is sufficient, in the most cases, to calculate $\Delta(t)$ at one point $t = -1$. The values of $t = -1$ and $t = -1$ distinguish the great majority of all 166 knots that can be drawn with less than 11 intersections on their projection [42], However, there are topologically different knots that have the same Alexander polynomials. In particular, $t = -1$ of a knot and its mirror image are identical, although very often such knots are topologically different. Thus, other methods are required if we want to distinguish among such knots. The problem can be solved by calculations of more powerful invariants, like the Jones polynomial [43], although this requires much more computer time [44]. In some cases calculation of Wr helps to distinguish between a knot and its mirror image [45].

3.5.2 Links

To define the topology of two chains, one can calculate the Alexander polynomial for two curves, Wr... [46]. The Alexander matrix for two chains is constructed similar to the matrix for one chain. Two contours are projected on an arbitrary chosen plane (see Fig. 3.12).

Renumbering of the generators, x_k, and the corresponding crossings starts in one contour and continues in the other. We denote the number of crossings on the first contour by M. Renumbering of the generators and crossings in the second contour starts from $M + 1$ and ends at N. Thus, the overpassing generator x_i for crossing k belongs to the first contour if $i \leq M$ and to the

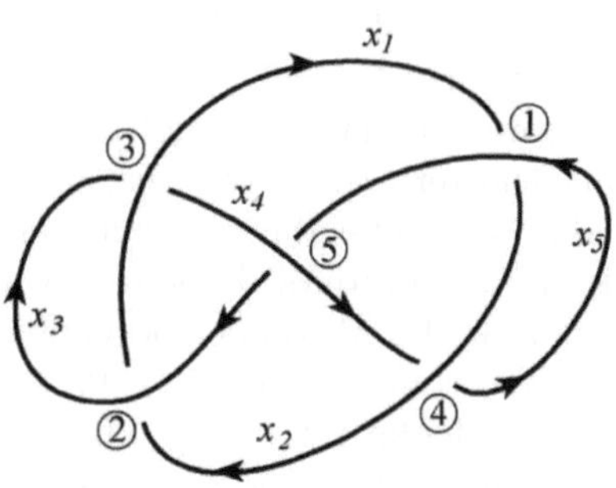

Fig. 3.12. Calculation of an Alexander polynomial for links. The *projection* is shown with all generators and crossing points

second contour if $i > M$. All elements of the Alexander matrix except a_{kk}, a_{kk+1} and a_{ki} are zero. The nonzero elements of the kth row are defined as follows:

(1) $k \le M,\ M > 1$
   (a) For $i = k$ or $i = k + 1$
      $a_{kk} = -1, a_{kk+1} = 1$ independent of the type of crossing;
   (b) For $i \ne k,\ i \ne k + 1,\ i \le M$
      $a_{kk} = 1, a_{kk+1} = -s; a_{ki} = s - 1$ for type I crossing,
      $a_{kk} = -s, a_{kk+1} = 1; a_{ki} = s - 1$ for type II crossing;
   (c) For $i > M$
      $a_{kk} = 1, a_{kk+1} = -t; a_{ki} = s - 1$ for type I crossing,
      $a_{kk} = -t, a_{kk+1} = 1; a_{ki} = s - 1$ for type II crossing;

(2) $k = M = 1;\ i > M$
   $a_{kk} = 1 - t, a_{ki} = s - 1$ independent of the type of crossing;

(3) $k > M;\ N > M + 1$
   (a) For $i = k$ or $i = k+1$
      $a_{kk} = -1, a_{kk+1} = 1$ independent of the type of underpass;
   (b) For $i \ne k,\ i \ne k + 1,\ i > M$
      $a_{kk} = 1, a_{kk+1} = -t; a_{ki} = t - 1$ for type I crossing,
      $a_{kk} = -t, a_{kk+1} = 1; a_{ki} = t - 1$ for type II crossing;
   (c) For $i \le M$
      $a_{kk} = 1, a_{kk+1} = -s; a_{ki} = t - 1$ for type I crossing,
      $a_{kk} = -s, a_{kk+1} = 1; a_{ki} = t - 1$ for type II crossing.

(4) $k = N;\ N = M+1,\ i \le M$
   $a_{kk} = 1 - s, a_{ki} = t - 1$ independent of the type of crossing.

These relationships hold under conditions that for $k = M$ one makes the substitution $(k + 1) \to 1$ and for $k = N$ the substitution $(k + 1) \to M + 1$. It is also assumed that there is at least one crossing in each contour formed by generators from different contours – in other case the contours are certainly unlinked.

Then one has to calculate a minor A_{kj} of order $N - 1$ of the Alexander matrix and divide it by $(s - 1)$ if $j \le M$ and by $(t - 1)$ if $j > M$. The resultant expression is multiplied by $(\pm t^{-m} s^{-n})$ (m and n are integers), so that the polynomial so obtained has no negative powers, and the positive powers are minimal, and the term with the largest total exponent must be positive. The polynomial $\Delta(s,t)$ defined in this manner is called *the Alexander polynomial* for the link of two contours. It is an invariant – it is rigorously proved in the knot theory that the Alexander polynomials $\Delta(s,t)$ coincide for equivalent links. For unlinked contours $\Delta(s,t) = 0$.

It is usually sufficient to calculate $\Delta(s,t) = 0$ for the majority of problems related with circular DNA molecules. The analysis shows that $\Delta(s,t) = 0$ is a more powerful topological invariant than the Gauss integral (which equals $\Delta(s,t) = 0$) [26]. It has different values for the simplest links and for unlinked

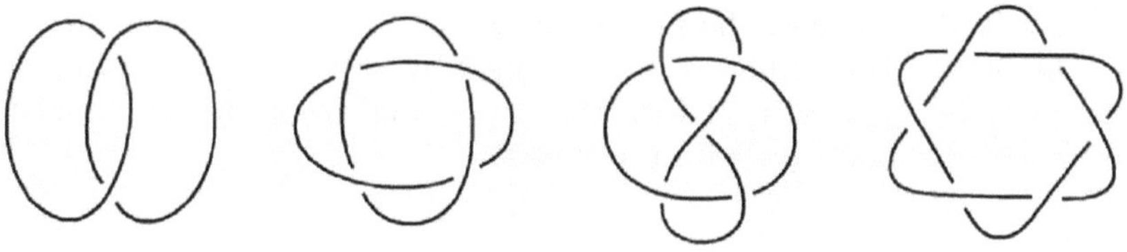

1 $s+t$ $(s-1)(t-1)$ s^2+t^2+st

Fig. 3.13. The simplest links and their Alexander polynomials, g$\Delta(s,t)$. All links that can be drawn with less than six intersections and one of three links with six intersections are shown. For an unlinked contour Δ(s,t) = 0

curves (for unlinked curves $\Delta(s,t) = 0$ equals 0)... [42]. In the most cases calculation of $\Delta(s,t) = 0$ also takes less computer time than that of the Gauss integral. Four simplest links and the corresponding $\Delta(s,t)$ are shown in Fig. 3.13. Complete table of links with less than 11 crossings can be found in [42].

Checking topology of a circular chain may be the most time-consuming part of the whole calculation. Therefore this part of the computer program deserves maximum attention in terms of its rationality. In this connection a procedure of reducing the order of the Alexander matrix before calculating it, by eliminating trivial intersections, is very useful (for details, see [26, 47, 48].

3.6 Calculation of Writhe

For many problems related with circular DNA one needs to calculate writhe, Wr, of a closed chain. It can be done by using definition of Wr through the Gauss integral (see [49], for example). For straight segments this integral may be presented in the form of a double sum of simple terms [50]. The method allows natural extension to the case of linear chains where the integral can serve as a measure of the chain chirality [32]. However, there are more convenient and efficient methods to calculate Wr [48, 50]. In the method suggested by Le Bret the total writhing value is presented in the form of two contributions, one of which is the directional writhing number, which can be calculated simultaneously with calculating the Alexander polynomial, virtually without additional computations. The directional writhing number in the z direction is the sum of +1 or −1 over all crossings. The sign of each term is determined by the type of crossing (Fig. 3.11). The second contribution is a sum over all elementary segments $\mathbf{r}_m$ of the chain:

$$\sum_m \{\arcsin[\sin \xi_m \sin(\varphi_{m+1} - \chi_m)] - \arcsin[\sin \xi_m \sin(\varphi_m - \chi_m)]\}/2\pi,$$

where φ_m is the angle of the projection of $\mathbf{r}_m$ on the x, y plane with the x axis. The perpendicular to the plane formed by vectors $\mathbf{r}_m$ and $\mathbf{r}_{m+1}$ makes the angle $\xi_m (0 \leq \xi_m \leq \pi/2)$ with the z axis, and its projection on the x, y plane makes the angle χ_m with the x axis.

3.7 Simulation Procedure

3.7.1 General Approach

The Metropolis–Monte Carlo procedure [51] is usually used for statistical sampling of the chain conformations. The procedure consists of consecutive steps that include displacements of certain parts of the chain. One may use different displacements in the procedure. Usually the displacement consists of the rotation of an arbitrary number of adjacent segments by a random angle within the interval $(-\varphi_0, +\varphi_0)$ around the straight line connecting two randomly chosen vertices, m_1 and m_2 (Fig. 3.14) [30].

The φ_0 value is adjusted during the simulation in such a way that about one-half of the steps would be successful. The rate of exchange between different conformations of a supercoiled chain can be further increased if φ_0 depends on the distance between vertices m_1 and m_2. To increase the rate of sampling in the simulation of supercoiled DNA we also introduced more complex type of motion, the so-called reptation motion [32]. Additional type of motion must be introduced for linear chains [14]. In general, one can introduce any type of motion to increase the rate of sampling as long as it does not interrupt the principle of microscopic equilibrium [51].

Whether the new conformation is accepted is determined by applying the rules of the procedure: (a) If the energy of the new conformation E_{new} is lower than the energy of the previous conformation, E_{old}, the new conformation is accepted, (b) If the energy of the new conformation is greater than the energy of the previous conformation, then the new conformation is accepted with the probability $p = \exp[(E_{\text{old}} - E_{\text{new}})/k_{\text{B}}T]$. The starting conformation is chosen arbitrarily.

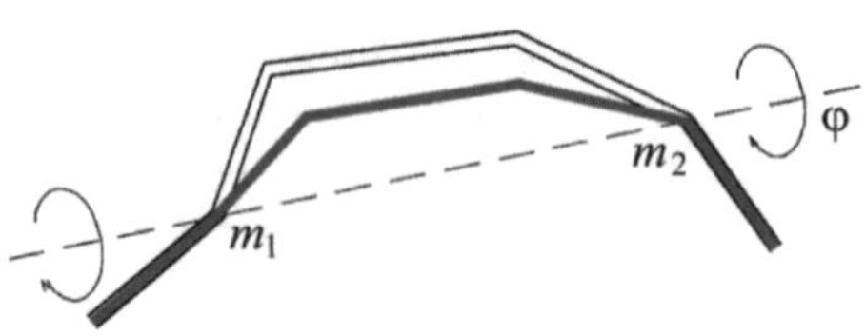

Fig. 3.14. The major type of displacement in the course of the Metropolis procedure described in the text

3.7.2 Simulation of DNA Conformations with Low Probability of Appearance

The Chain of Conditional Probabilities

Although modern computers allow one to perform up to 10^9 moves for a model chain corresponding to DNA molecule a few kb in length, this may not be sufficient to evaluate probabilities of some rare conformations with a reasonable statistical error. This can be the case for juxtapositions of specific sites or DNA ends, for example. Two methods have been developed to overcome this problem.

The first method enables one to calculate values of j-factors for short DNA fragments, about 200 base pairs, when these values are less than 10^{-8} M and direct Monte Carlo simulation is inefficient. Suppose we want to estimate $P(r_0)$, the probability of the conformations with end-to-end distance, r, less than a small value r_0. We choose a sequence of distances $r_0 < r_1 < \cdots < r_n$, where r_n is larger or equal to the chain contour length. Let $P(r_i)$ be the probability of conformations with $r < r_i$. We can also define the conditional probabilities, $P(r_i|r_{i+1})$, of conformations with $r < r_i$ in the subset of conformations with $r < r_{i+1}$. Since $P(r_i) = P(r_i|r_{i+1})P(r_{i+1})$ and $P(r_n) = 1$, the value of $P(r_0)$ can be found as

$$P(r_0) = \prod_{i=0}^{n-1} P(r_i|r_{i+1}). \tag{3.13}$$

The sequence of distances $r_0 < r_1 < \cdots < r_n$ can be chosen so that all $P(r_i|r_{i+1})$ values are relatively large. This can be always achieved since $P(r_i|r_{i+1})$ approaches 1 when r_{i+1} approaches r_i. The large values of $P(r_i|r_{i+1})$ can be efficiently and accurately calculated by the Metropolis procedure. Each $P(r_i|r_{i+1})$ is calculated as the fraction of the conformations with $r < r_i$ in the subset of equilibrium conformations with $r < r_{i+1}$. These subsets are generated in the Monte Carlo procedure by rejecting any trial conformation with $r > r_{i+1}$. The values $P(r_i|r_{i+1})$ are calculated sequentially from $P(r_0|r_1)$ to $P(r_{n-1}|r_n)$. The starting conformation for each subset is the last conformation from the previous subset. The calculation of $P(r_0|r_1)$ is started from a conformation with $r = 0$. The estimation shows that the best efficiency in estimating $P(r_0)$ is achieved when the values of $P(r_i|r_{i+1})$ are close to 0.2. Using this approach, one can speed up the computations by a few orders of magnitude compared to the direct Monte Carlo procedure [39, 52].

Umbrella Method

The umbrella method [53] addresses calculation of conformational distributions under the condition that specific sites are juxtaposed in a proper orientation. It is based on introducing an artificial potential, $U(\mathbf{x})$, where $\mathbf{x}$

refers to the coordinates that define the mutual geometry of the specific sites. Although $U(\mathbf{x})$ can greatly increase the probability of the site juxtaposition, it does not disturb the conditional distribution since $U(\mathbf{x})$ has the same value for all conformations where the sites are juxtaposed ($\mathbf{x} = \mathbf{x}_0$). Indeed, the statistical weights of all conformations with the juxtaposed sites will be multiplied by the same factor, $\exp(-U(\mathbf{x}_0)/kT)$. The efficiency of the approach, which is called Umbrella method, depends on proper choice of the potential $U(\mathbf{x})$. The method was recently applied for the simulation of the juxtaposition of specific sites in supercoiled DNA [54].

Acknowledgment

The author thanks the Max Plank Institute of Physics of the Complex Systems for their hospitality during the workshop on "Topology in Condensed Matter" Dresden 2002.

References

1. P.J. Hagerman, Biopolymers **20**, 1503–1535 (1981)
2. P.J. Hagerman, B.H. Zimm, Biopolymers **20**, 1481–1502 (1981)
3. V.V. Rybenkov, A.V. Vologoskii, N.R. Cozzarelli, J. Mol. Biol. **267**, 299–311 (1997c)
4. P.J. Hagerman, Annu. Rev. Biochem. **59**, 755–781 (1990)
5. S.D. Levene, D.M. Crothers, J. Mol. Biol. **189**, 73–83 (1986)
6. W.H. Taylor, P.J. Hagerman, J. Mol. Biol. **212**, 363–376 (1990)
7. M. Vologodskaia, A. Vologodskii, J. Mol. Biol. **317**, 205–213 (2002)
8. K.V. Klenin, A.V. Vologodskii, V.V. Anshelevich, A.M. Dykhne, M.D. Frank-Kamenetskii, J. Biomol. Struct. Dyn. **5**, 1173–1185 (1988)
9. K.V. Klenin, A.V. Vologodskii, V.V. Anshelevich, V.Y. Klisko, A.M. Dykhne, M.D. Frank-Kamenetskii, J. Biomol. Struct. Dyn. **6**, 707–714 (1989)
10. V.V. Rybenkov, N.R. Cozzarelli, A.V. Vologodskii, Proc. Natl. Acad. Sci. USA **90**, 5307–5311 (1993)
11. V.V. Rybenkov, A.V. Vologodskii, N.R. Cozzarelli, J. Mol. Biol. **267**, 312–323 (1997b)
12. S.Y. Shaw, J.C. Wang, Science **260**, 533–536 (1993)
13. A.V. Vologodskii, N.R. Cozzarelli, J. Mol. Biol. **232**, 1130–1140 (1993)
14. A.V. Vologodskii, Macromolecules **27**, 5623–5625 (1994)
15. A.V. Vologodskii, J.F. Marko, Biophys. J. **73**, 123–132 (1997)
16. J.A. Gebe, J.J. Delrow, P.J. Heath, B.S. Fujimoto, D.W. Stewart, J.M. Schurr, J. Mol. Biol. **262**, 105–128 (1996)
17. M. Hammermann, N. Brun, K.V. Klenin, R. May, K. Toth, J. Langowski, Biophys. J. **75**, 3057–3063 (1998)
18. M. Hammermann, C. Stainmaier, H. Merlitz, U. Kapp, W. Waldeck, G. Chirico, J. Langowski, Biophys. J **73**, 2674–2687 (1997)
19. K. Klenin, M. Hammermann, J. Langowski, Macromolecules **33**, 1459–1466 (2000)

20. S. Allison, R. Austin, M. Hogan, J. Chem. Phys. **90**, 3843–3854 (1989)
21. S.A. Allison, Macromolecules **19**, 118–124 (1986)
22. G. Chirico, J. Langowski, Biopolymers **34**, 415–433 (1994)
23. J. Huang, T. Schlick, T. Vologodskii, Proc. Natl. Acad. Sci. USA **98**, 968–973 (2001)
24. J. Vinograd, J. Lebowitz, R. Radloff, R. Watson, P. Laipis, Proc. Natl Acad. Sci. USA **53**, 1104–1111 (1965)
25. W.R. Bauer, F.H.C. Crick, J.H. White, Sci. Am. **243**, 100–113 (1980)
26. M.D. Frank-Kamenetskii, A.V. Vologodskii, Sov. Phys.-Usp. **24**, 679–696 (1981)
27. F.B. Fuller, Proc. Natl. Acad. Sci. USA **68**, 815–819 (1971)
28. F.B. Fuller, Proc. Natl. Acad. Sci. USA **75**, 3557–3561 (1978)
29. A.V. Vologodskii, in *On-Line Biophysics Textbook*, ed. by V. Bloomfield. http://www.biophysics.org/btol/supramol.html#13
30. M.D. Frank-Kamenetskii, A.V. Lukashin, V.V. Anshelevich, A.V. Vologodskii, J. Biomol. Struct. Dyn. **2**, 1005–1012 (1985)
31. A.V. Vologodskii, N.R. Cozzarelli, Biopolymers **35**, 289–296 (1995)
32. A.V. Vologodskii, S.D. Levene, K.V. Klenin, M.D. Frank-Kamenetskii, N.R. Cozzarelli, J. Mol. Biol. **227**, 1224–1243 (1992)
33. D. Stigter, Biopolymers **16**, 1435–1448 (1977)
34. P.J. Hagerman, Ann. Rev. Biophys. Biophys. Chem. **17**, 265–286 (1988)
35. D.S. Horowitz, J.C. Wang, J. Mol. Biol. **173**, 75–91 (1984)
36. A.A. Brian, H.L. Frisch, L.S. Lerman, Biopolymers **20**, 1305–1328 (1981)
37. V.V. Rybenkov, A.V. Vologodskii, N.R. Cozzarelli, Nucl. Acids Res. **25**, 1412–1418 (1997a)
38. V. Katritch, A. Vologodskii, Biophys. J. **72**, 1070–1079 (1997)
39. A.A. Podtelezhnikov, C. Mao, N.C. Seeman, A.V. Vologodskii, Biophys. J. **79**, 2692–2704 (2000)
40. A.V. Vologodskii, A.V. Lukashin, M.D. Frank-Kamenetskii, V.V. Anshelevich, Sov. Phys. JETP **39**, 1059–1063 (1974)
41. R.H. Crowell, R.H. Fox, *Introduction to Knot Theory* (Springer-Verlag, New York, Heidelberg, Berlin, 1963)
42. D. Rolfsen, *Knots and Links* (Publish or Perish, Inc., Berkeley, CA, 1976)
43. K. Murasugi, *Knot Theory and its Applications* (Birkhauser, Boston, 1996)
44. A.A. Podtelezhnikov, N.R. Cozzarelli, A.V. Vologodskii, Proc. Natl. Acad. Sci. USA **96**, 12974–12979 (1999)
45. K. Klenin, J. Langowski, A. Vologodskii, J. Mol. Biol. **320**, 359–367 (2002)
46. A.V. Vologodskii, A.V. Lukashin, M.D. Frank-Kamenetskii, Sov. Phys. JETP **40**, 932–936 (1975)
47. J. Des Cloizeaux, M.L. Metha, J. Physique **40**, 665–670 (1979)
48. M. Le Bret, Biopolymers **19**, 619–637 (1980)
49. A.V. Vologodskii, *Topology and Physics of Circular DNA* (CRC Press, Boca Roton, 1992)
50. K. Klenin, J. Langowski, Biopolymers **54**, 307–317 (2000)
51. N. Metropolis, A.W. Rosenbluth, M.N. Rosenbluth, A.H. Teller, E. Teller, J. Chem. Phys. **21**, 1087–1092 (1953)
52. A.A. Podtelezhnikov, A.V. Vologodskii, Macromolecules **33**, 2767–2771 (2000)
53. J.A. McCammon, S.C. Harvey, *Dynamics of Proteins and Nucleic Acids* (Cambridge University Press, Cambridge, UK, 1987)
54. I. Grainge, S. Pathania, A. Vologodskii, R. Harshey, M. Jayaram. J. Mol. Biol. (2002)

4

Dynamics of DNA Supercoiling

A. Gabibov, E. Yakubovskaya, M. Lukin, P. Favorov, A. Reshetnyak, and M. Monastyrsky

Summary. A catalytic turnover of supercoiled DNA (scDNA) transformation mediated by topoisomerases leads to the changes of the linking number (Lk) of the polymeric substrate by 1 or 2. While a substrate of the topoisomerisation reaction is chemically identical to its product, even single catalytic event results in the quantum leap in the scDNA topology. A continuous non-disturbing assay for measurement of kinetics of the scDNA topoisomerisation was lacking. The intrinsic connections of DNA topology, its hydrodynamics and optical anisotropy, studied in this chapter allowed the use of flow linear dichroism technique (FLD) for continuous monitoring of scDNA topoisomerisation reaction. This approach permits studying the kinetics of DNA transformation catalysed by eukaryotic topoisomerases I and II, mechanistic properties of these enzymes and their interactions with anti-cancer drugs.

Keywords: scDNA transformations, DNA hydrodynamics, linking number (Lk), topoisomerases, flow linear dichroism, anti-cancer drugs

4.1 Introduction

The dynamics of scDNA transformations is a key point for understanding the numerous processes that take place in the living cell [1]. Changes of DNA topology are vital during replication, transcription, recombination, chromosome condensation and segregation. From the topological point of view DNA can be represented by a closed ribbon [2, 3]. Studies of the dynamical aspects of DNA topology are closely connected with the design of the adequate mathematical description of DNA polymeric molecule as well as the methods of monitoring of its properties [4, 5]. The main topological changes of scDNA in the cells are catalysed by DNA-specific enzymes, topoisomerases, types I and II, which induce single and double nicks in DNA strains. This leads to changes of the linking number (Lk) of the polymer substrate by 1 or 2. The edges of "closed ribbon" cannot be regarded as "intact" during the catalytic reaction and DNA molecule can be preferably represented by a "ladder" instead of "closed ribbon". Here by the term *ladder* we assume the lattice surface where the points (nucleotides) are located on the boundary curves and connected by edges (phosphodiester bonds). These edges may be disrupted. From the chemical point of view, the substrates and products during the DNA topoisomerisation are identical and the catalytic events result in small topological changes. The product from the previous single turnover will play the role of the substrate in the next stage. So the ensemble of topoisomers exists at each step of the reaction. Even a single catalytic event results in the quantum leap in the scDNA topology. This allows study of most of biocatalytic problems of DNA supercoiling within the frame of the problems of DNA topology, DNA hydrodynamics and statistics of biopolymers. The general scheme of enzyme-mediated topological transformations of DNA is displayed in Fig. 4.1.

DNA molecule runs a whole sequence of the states, from a supercoiled one to the relaxed one. Moreover, either of those steps could be reversible.

In this connection, to describe kinetics of DNA relaxation correctly it is necessary:

(a) To possess information about the instant concentrations of either of the topoisomers involved in the reaction.
(b) To identify integral index as a function of these concentrations and reflecting the reaction turnover.

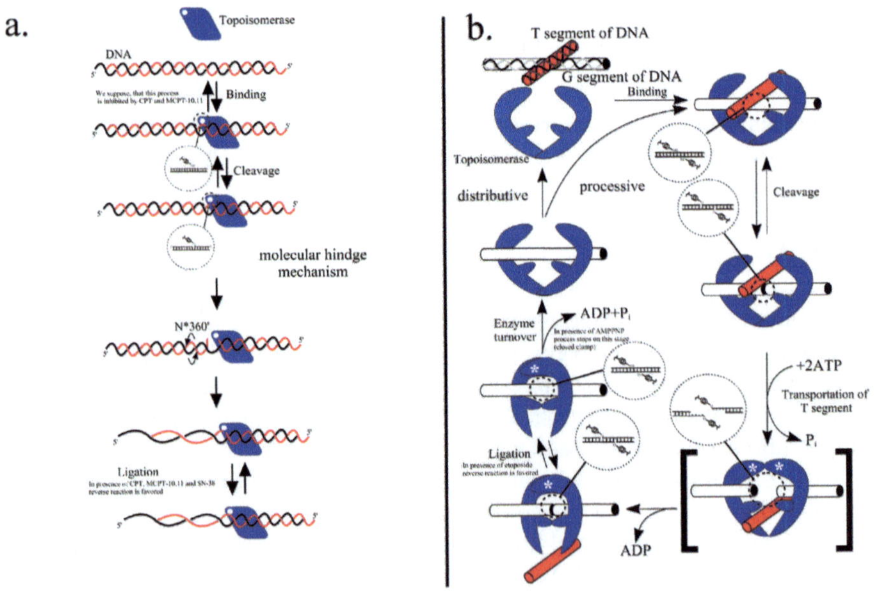

Fig. 4.1. Mechanisms of enzyme-mediated DNA relaxation. (**a**) Topoisomerase I-3′. (**b**) Topoisomerase II. Two possible ways of topoisomerase action, distributive and processive kinetic schemes (see text for details)

Topoisomerases are much smaller than the polymeric substrates in vivo and their interaction with DNA is restricted within a small area (1–2 turns of DNA helix). The enzyme cannot recognise a conformation of a whole DNA molecule and proceeds only with a short segment of double helix (or with single crossover formed by the helix itself).

The Gibbs energy function describing conformational transitions of scDNA depends on the superhelical density and determines the direction of topological transformation and its rate [6]:

$$\Delta\Delta G = BRT\sigma, \tag{4.1}$$

where σ is the superhelical density ($\sigma = \Delta\mathrm{Lk}/k_0$) and B is the rigidity index and $\mathrm{Lk}_{\mathrm{fin}}$, $\mathrm{Lk}_{\mathrm{init}}$ are the final and initial linking numbers. For topoisomerases I and II Lk changes by 1 and 2 correspondingly, during every step of topoisomerisation (Fig. 4.1). $\Delta\Delta$G becomes proportional to σ, which is independent of DNA length.

The above-mentioned statements mean that the parameter reflecting a degree of DNA conversion should be a parameter that has no relation to the concentration of either of the topoisomers averaged to the topological state of DNA under the study $\langle\sigma\rangle$.

The introduction of $\langle\sigma\rangle$ function provides DNA topoisomerisation studies with additional opportunities of quantitative estimations. The intrinsic links between $\langle\sigma\rangle$ and physical property of the DNA molecule, namely linear dichroism (LD), may provide a serious experimental basis for the field.

4.2 Theory

A continuous method of statistical mechanics of biopolymers was successfully applied to the theoretical description of the processes of DNA supercoiling [7, 8]. For the DNA described by the "closed ribbon" model the Calugareanu–Fuller–White formula can be written:

$$\mathrm{Lk} = \mathrm{Tw} + \mathrm{Wr}, \tag{4.2}$$

where Lk is the Gauss linking number, Tw is the twist and Wr is the writhing number [3, 9–11]. All definitions in (4.2) are well known topologically. Let us regard γ as a closed smooth curve embedded in R^3 (Euclidean space). Then υ is a normal vector field on γ. Let us assume that the magnitude of a vector $\upsilon(\mathrm{t})$ is so small that $\upsilon(\mathrm{t})$ intersects γ only at one point. The endpoint of υ sweeps a curve γ_υ, which inherits the orientation of γ, while υ itself inherits a strip embedded in R^3. Then the twist of υ can be defined as follows:

$$\mathrm{Tw} = \frac{1}{2\pi}\int_\gamma \upsilon^\perp \mathrm{d}\upsilon, \tag{4.3}$$

where vector $\upsilon^{\perp}$ is in the frame $(t, \upsilon g\ \upsilon^{\perp})$, a right-hand system, and t is the unit tangent vector to the curve γ. The twist of the curve is a continuous quantity. The writhing number is the integral:

$$\mathrm{Wr} = \frac{1}{4\pi} \int_{\gamma} \int_{\gamma} \frac{([\mathrm{d}r_1 \mathrm{d}r_2], r_1 - r_2)}{(r_1 - r_2)^3}, \tag{4.4}$$

Wr is determined directly by the curve γ and is merely evaluated in the experiment. It is also a continuous quantity.

The left side of formula (4.1) is the Gauss linking number:

$$\mathrm{Lk} = \frac{1}{4\pi} \iint_{\gamma\gamma_{\upsilon}} \frac{([\mathrm{d}r_1 \mathrm{d}r_2], r_1 - r_2)}{(r_1 - r_2)^3}, \tag{4.5}$$

where r is the radius vector of the curve's point, and [ ] and ( ) are the vector and scalar products, respectively. The two main questions to be considered when the real-time kinetics of scDNA biocatalytic conversion is studied are: (1) how the equations, suitable for the "closed ribbon" DNA model, could be applied to the more realistic "ladder" model and (2) how to estimate the changes of the ensemble of topoisomers during the reaction time course.

Let us assume that the process of removal of supercoils takes place exclusively by cutting the edges of the DNA ribbon, twisting and sewing the band. In this case, formula (4.2) can be rearranged as:

$$(\mathrm{Lk} - q) = \widetilde{\mathrm{Tw}} + \widetilde{\mathrm{Wr}}. \tag{4.6}$$

Here q is the number of cutting–twisting–sewing events, and Tw and Wr are modified parameters of formula (4.2).

As (4.4) is purely topological, it is also valid for the ladder DNA model, and all parameters can be described in terms of simplicial divisions [12]. Obviously, we deal with the ensemble of cuts and sews, which requires the averaging over all states:

$$\langle(\mathrm{Lk{-}q})\rangle = \left\langle\widetilde{\mathrm{Tw}}\right\rangle + \left\langle(\widetilde{\mathrm{Wr}}\right\rangle. \tag{4.7}$$

Finally this explains the applicability of topological approach for the real-time course of the scDNA relaxation, caused by enzymatic activity of topoisomerases. It can be shown that intrinsic hydrodynamic behaviour of DNA is closely related with this process. In fact the tensor parameter of order, biaxial in general, can be reduced for the free-rotating DNA to the uniaxial one. Then, the parameter of order as in the case of nematic liquid crystal [13] can be described as follows:

$$A_{ik} = A_0(n_i n_k - 1/3\delta_{ik}), \tag{4.8}$$

where $n = (n_1, n_2, n_3)$ is the unit vector, and δ_{ik} is the Kronecker delta function. Topological characteristics of DNA are connected with hydrodynamic equations and this equation can be presented as:

$$\frac{\partial n}{\partial t} = \{H, n\}, \tag{4.9}$$

where H is the Hamiltonian of the system determined as the functional of the energy density $n = \int E \, \mathrm{d}^3 r$, where E depends on the states of the system and $\{H, n\}$ is the Poisson brackets. On the other hand, vector $\mathbf{n}$ in the case of appropriate boundary conditions can be connected with the topological invariant Lk

$$\mathrm{Lk}(\gamma, \gamma_v) = \int (\mathbf{n}, \ \mathrm{curl}\,\mathbf{n}) \mathrm{d}V. \tag{4.10}$$

Here $\mathrm{d}V$ is the volume element of the ball, embracing the band (γ, γ_ν).

The combination of formulas (4.5) and (4.7) gives the final equation

$$\left\langle \int (\mathbf{n}, \ \mathrm{curl}\,\mathbf{n}) \mathrm{d}V - q \right\rangle = \left\langle \widetilde{\mathrm{Tw}} \right\rangle + \left\langle \widetilde{\mathrm{Wr}} \right\rangle. \tag{4.11}$$

All these facts display the interrelation between the intrinsic properties of DNA molecule, i.e. optical anisotropy, hydrodynamics and topology. This conclusion allows us to study the changes of DNA topology by monitoring the optical properties of oriented DNA molecules.

4.2.1 Flow Linear Dichroism and Dynamics of DNA Supercoiling

The time-dependent topological transformations of supercoiled DNA can be visualised using electrophoretic analysis of reaction mixture at different stages of the process. However, this technique cannot provide instant and non-disturbing quantitative kinetic analysis of reaction. A recently published precise approach based on the immobilisation of a single DNA molecule [14] has demanded a state-of-the-art technique and cannot be utilised as a routine quantitative monitoring technique. Previously, we had first applied the FLD method for the kinetic analysis of different nuclease reactions, using flow-oriented supercoiled DNA molecule as a substrate [15–19]. This method is based on the fact that oriented DNA possesses the property of optical anisotropy [20, 21]. When the polymer molecule changes its topology, by low-molecular weight effectors (as benzpyrene) [22] or biocatalysts [16, 23] this instantly affects its hydrodynamics. This has an influence on the orientation O and optic factors S of DNA molecule, and leads to the alteration of the of linear dichroism value. The principle of the method is displayed in Fig. 4.2.

The value of linear dichroism ρ is defined as the ratio between ΔA, difference in light absorption polarised parallel and perpendicular to the DNA orientation axis, and A, optical density of a sample in non-polarised light [21]. For any sample containing optically anisotropic molecule, ρ is a product of orientation O and optic factors S:

$$\rho = SO, \tag{4.12}$$

where the first one is the degree of alignment of the molecule along the orientation axis, and the second one is the intrinsic optical anisotropy of the molecules.

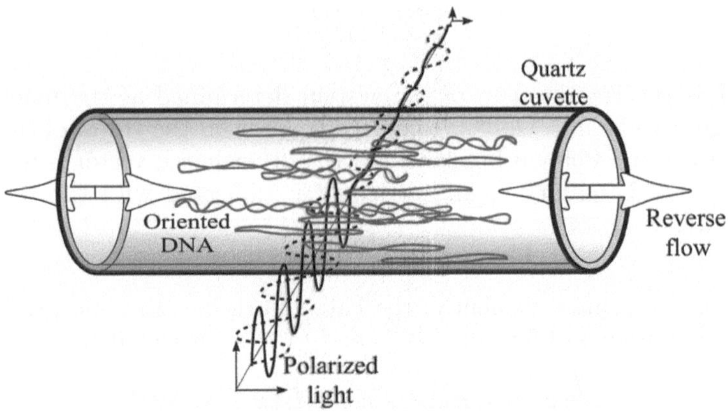

Fig. 4.2. Principle of the flow linear dichroism (FLD) method. Experimental conditions: DNA polymers are oriented in reversible flow of reaction mixture. Combined vector of orientation of chromophoric groups of DNA is then different from zero, its value and sign depending on topological state of molecules and their orientation ability. Measured value is the ratio between the difference of the optical densities of the oriented sample for two perpendicularly polarised rays and the optical density for non-polarised light ($\Delta A/A$). Absolute values of reduced linear dichroism (ρ) were measured by using a JASCO J500C spectropolarimeter equipped with an achromatic quarter-wavelength prism at 260 nm. Plasmid DNA was oriented by the flow gradient provided by pumping the solution through the flow cell using the reciprocal pump, designed and constructed by V.L. Makarov (Engelhardt Institute of Molecular Biology, Russian Academy of Sciences) [24]. The cell volume was 200 µl; optical path length was 2 mm; the frequency of reciprocal pump was 100 rpm; an average flow gradient in the cell was 3,000 s^{-1}

An scDNA can be relatively adequate rendered as a rigid rod formed by a double helix winded into the superhelix, which could further form superhelices of higher order. For such molecules ρ is defined as

$$\rho = 1/2(3\langle\cos^2 \Theta\rangle - 1)(1/2(3\langle\cos^2 \beta_i\rangle - 1)), \tag{4.13}$$

where Θ is the angle between the main axis of DNA molecule and axis of orientation of the molecules. The first coefficient is thus an orientation factor, which under the condition of shear flow is determined by hydrodynamic properties of the molecule, and is dependent on the form of DNA molecule and its rigidity. The rest of coefficients describes the intrinsic optical anisotropy of scDNA. Parameter $\beta_i = tg^{-1}(2\pi R_i/P_i)$, where R_i and P_i are radius and pitch of the ith order superhelix [21].

It seems to be clear that all variables (Θ, R_i and P_i) in this equation are strongly dependent on $\langle\sigma\rangle$. As an example, sc circular DNA can be better oriented than the relaxed circles of the same counter length. However, formation of the supercoiled DNA is accompanied by shortening the radius and increasing the superhelix density (R_i is reduced and P_i is increased), which causes a decrease of the optic anisotropy of the molecule. Thus the dependence of optic anisotropy of circular DNA and its topology should probably have a bell-shaped form with a maximum in a point corresponding to the completely relaxed DNA. It should reach the minimum at moderate $|\sigma|$ and then should rise again.

This assumption was confirmed by titration experiments (circular DNA by intercalators, proflavine, ethidium bromide, etc.), displaying the bell-shaped dependence of $\rho(\sigma)$ [22, 25]. Figrue 4.3 illustrates typical results of ethidium bromide titration of the negatively supercoiled plasmids. All the curves are characterised by the maximum corresponding to the fully relaxed state (point A). The presence of positive ($\sigma > 0$) or negative supercoils results in decrease of ρ (AD and AB segments of the curves). Further increase of $|\sigma|$ results in increasing of ρ again (segment BC). The latter probably is a consequence of the rising alignment ability of the DNA (in agreement with the previous theoretical prediction).

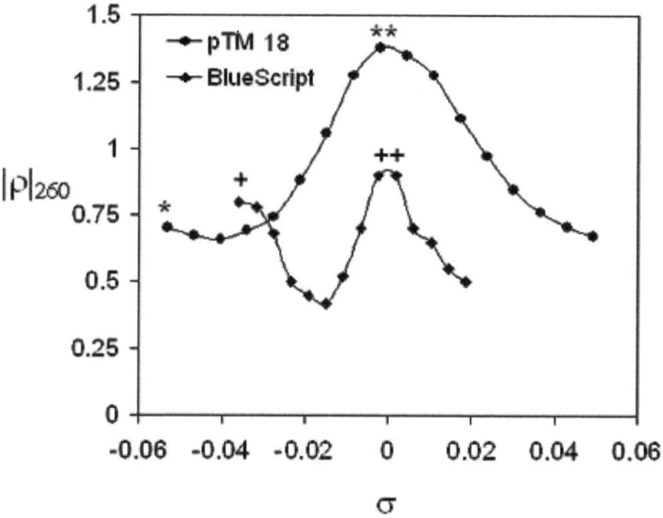

Fig. 4.3. The dependence of the FLD signal (ρ_{260}) on the density of supercoiling for plasmids pTM 18 and BlueScript. Linear dichroism of the sample is measured as relative value ρsg that is equal to zero for the blank (reaction buffer) and to 1 for the relaxed plasmid sample. The FLD measurements were performed in 200 µl of buffer (20 mM Tris–HCl (pH 7.5), 100 mM NaCl, 0.5 mM EDTA) containing 5 µg pTM 18 or 5 µg BlueScript at increasing concentration of ethidium bromide (1-µl aliquots of 6×10^{-5} M were added). The value of the superhelix density (σ) was calculated taking the unwinding angle value for ethidium bromide as 26° per intercalated ethidium ion [25]. The starting value of σ appeared to be −0.036 and −0.055 for BlueScript and pTM 18, respectively

Generally, the shape of $\rho(\sigma)$ curves is similar for a number of tested plasmids. However, quantitatively, the parameters describing each curve revealed a significant dependence upon specific character of the plasmids under study. The two plasmids illustrating the alternative FLD signal dependencies have been chosen. A pTM plasmid, while possessing very short BC segment, revealed the highest amplitude of the LD signal in the course of the transition from a supercoiled state to the relaxed one. BlueScript plasmid titration curve, on the contrary, contains a substantial BC segment. We observed the similar LD values for both supercoiled and relaxed states of this plasmid, but LD values for the transient topoisomers were much lower. Both the plasmids were applied as a tool for the topoisomerase I studies.

4.2.2 Mechanisms of Biocatalytic DNA Relaxation

For modelling enzyme-mediated DNA topoisomerisation the following assumptions were made:

At any time, ti, the reaction mixture contains unbound enzyme and plasmid in different topological states S^n (where n is the number of superturns). The enzyme and either of the DNA topoisomers can be bound to each other in an equilibrium way, to form a non-covalent complex ES^n, where both the DNA strands have no nicks. This complex is not capable of proceeding to topoisomerisation. The biocatalytic productivity requires the transfer of ES^n complex into the covalent complex (ES^n). Here one of the DNA strains is nicked and the enzyme is covalently bound to the 3′-end. The ES^n complex is capable of chaning its topology (the driving force of this process is the tension of supercoiled DNA molecule) to form ES^{n+1} or ES^{n-1}. The reaction is fully reversible and (ES^n) may take part in the back reaction. This mechanism is displayed in:

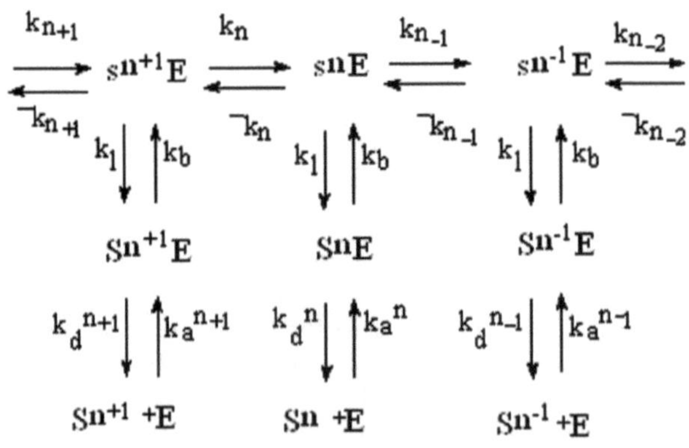

Scheme 4.1

This mechanistic representation allows to make quantitative estimations. For the calculations, we need to get values of the appropriate kinetic constants.

However a full set of the constants regarding human topoisomerases is not available. In this connection, as regarding the values of kinetic constant derived from (4.1) the following assumption have been made:

1. Since in the given reaction the driving force of the topoisomerisation depends on the tension of DNA that is proportional to the square of DNA density of superhelix [6], the dependence of the constants of topoisomerisation $\sim k^n$ and k^n over the density of the superturns looks as:

$$k_n = k_r \exp((Bn + 0.5)/RT), \tag{4.14}$$

$$\sim k_n = k_r \exp((-Bn + 0.5)/RT). \tag{4.15}$$

2. It is clear that $k_r = \sim k_{(0)} = k_{(0)}$, i.e. that k_r is the topoisomerisation rate constant of the covalent complex of the fully relaxed DNA with a topoisomerase. In accordance with the preceding evaluations, k_r value is rather high, so when calculated some of the k_r values ranging between 3 and $60\,\mathrm{s}^{-1}$ were applied.
3. The cleavage and ligation rate constants were taken as equal to 0.1 and 1 min^{-1}, respectively. These values were estimated for the vaccine virus topoisomerase using shot oligonucleotide substrates. Taking into account the low rigidity of DNA, it seems to be plausible that those constants reveal independence of the DNA topology [26].
4. The value of the association rate constant k_{ass}, was equal to $1,000\ \mathrm{s}^{-1}$, which had to be higher than k_{diss}, k_1, $k_{c1.}$
5. The value of the dissociation rate constant was taken either as independent of the topological state of DNA or it grew along with the increase of $|\sigma|$.

For modelling a reaction of topoisomerase II running in accordance with a "cleavage-changing linking number by 2-ligation" the following kinetic scheme was proposed:

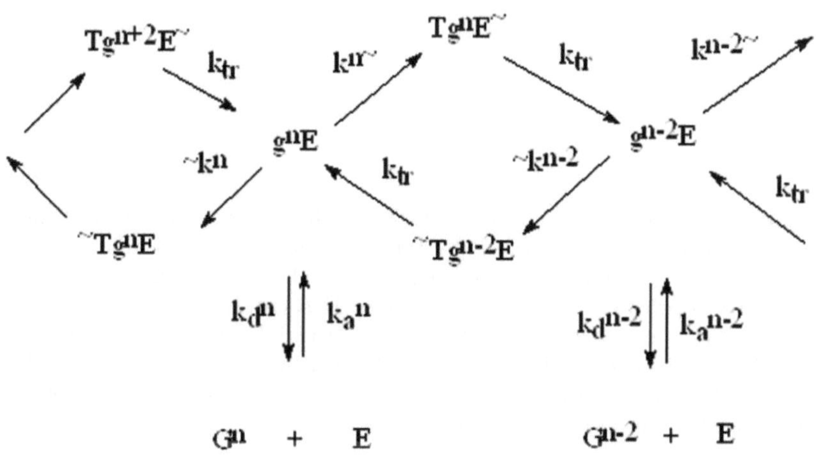

Scheme 4.2

This scheme reflects the most probable mechanism of top II action [27, 28]. Free enzyme associates with the first segment of DNA containing n superturns. As it will serve as a Gate segment at the next step of reaction we denote it as G^n. The formation and the dissociation of the g^nE complex are described with association and dissociation constants (k_a^n and k_d^n, respectively), which, in general, could depend on σ. Then the second segment of DNA, the T-segment, binds to the g^nE complex giving two types of ternary complexes, $Tg^nE^{\sim}$ and $^{\sim}Tg^nE$ (Fig. 4.4a). Since T- and G-segments practically coincide after the formation of these complexes, effective change of the DNA linking number (by approximately ± 1) takes place. These complexes have different constants of

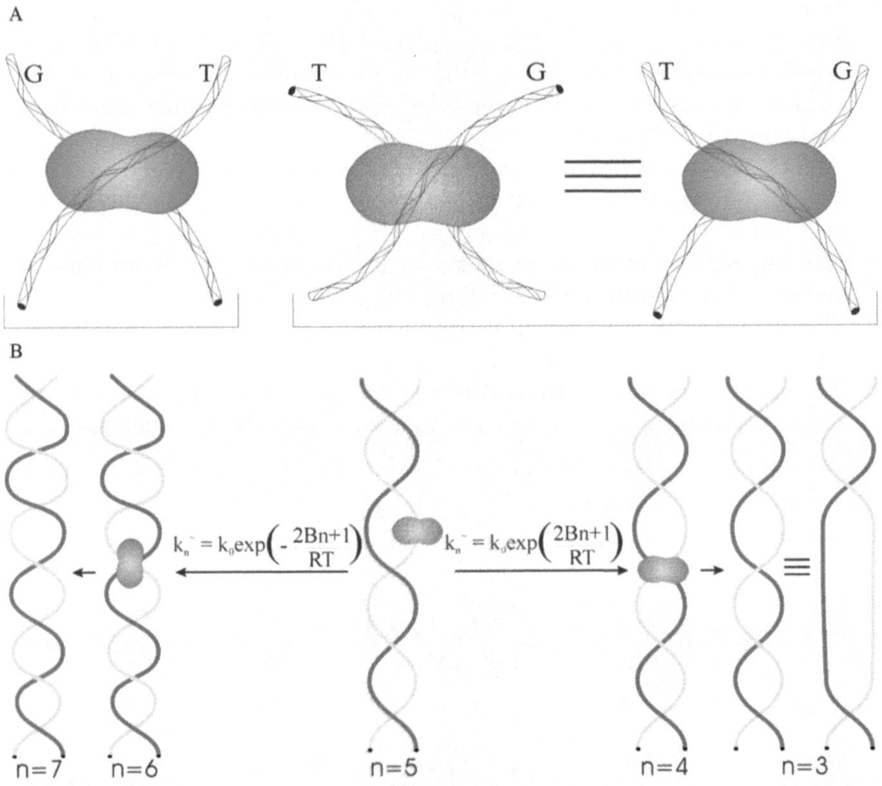

Fig. 4.4. Analysis of the complexes of scDNA with topoisomerase II. (**a**) Schematic representation of complexes formed by topoisomerase II with G and T DNA segments. A sign of the linking number depends on the superhelix symmetry of the complex. (**b**) Topological transformations of scDNA catalysed by topoisomerase II. Changes of the type of ternary complex effect on the linking number (approximately by ± 1). The Gibbs energy change is characterised by factor $4Bn/RT$. The kinetic constants reflecting the transformations between $(n+2)$ and $(n-2)$ topoisomers are described by $\exp(-4Bn/RT)$ function

formation ($k_n^{\sim} = k_0 \exp((2Bn+1)/RT)$ and $^{\sim}k_n = k_0 \exp(-(2Bn+1)/RT)$) (Fig. 4.4b). After formation of $Tg^nE^{\sim}$ and $^{\sim}Tg^nE$, followed by reversible cleavage of the gate segment, the direction of topoisomerisation (i.e. sign of the linking number change) is strictly defined by the symmetry of complex formed. The latter process uses ATP hydrolysis as an external source of energy. The rate of this step is assumed to be topologically independent (the k_t value is proportional to the DNA-induced ATP hydrolysis constant).

A system of differential equations that describes concentration change of either the reaction mixture components can be written for both cases. Although it could not be solved analytically, the numerical solution is available. Then the $\langle\sigma\rangle$ and $\langle\rho\rangle$ values at any time can be calculated according to the formula:

$$\langle\sigma\rangle = \Sigma g\sigma g T_n / \Sigma T_n, \tag{4.16}$$

and

$$\langle\rho\rangle = \Sigma(g\rho o \sigma_n) T_n, \tag{4.17}$$

where $\rho(\sigma)$ – as an empiric function calculated using the dependencies shown in Fig. 4.3 and T_n denotes the concentration of DNA with n superturns.

The two ultimate cases of $\rho(t)$ during DNA relaxation mediated by topoisomerases were obtained in the couarse of the numerical simulation of the relaxation process. This evaluation is dependent on the enzyme type and reaction conditions:

1. After binding of the enzyme to supercoiled DNA, a fast and virtually complete relaxation of DNA takes place. Dissociation of resulting enzyme – relaxed DNA complex occurs. This process possibly takes place in the case of eukaryotic topoisomerase I [29] and topoisomerases II under optimal conditions. If excess DNA is present in reaction mixture, only initial (supercoiled) and final (relaxed) topoisomers exist during the reaction course (Fig. 4.5), and (4.13) can be rearranged in (4.14):

$$\rho(t) = (1 - \lambda(t))\rho_{\text{init}} + \lambda(t)\rho_{\text{lot}}, \tag{4.18}$$

   where $\lambda(t)$ is the level of completion of the reaction.
2. For enzymes that reversibly nick DNA and change the ΔLk strictly by 2 (for euckaryotic topoisomerases II, Scheme 2) the ligation of the gap formed occurs in the next step. Significant amounts of intermediate topoisomers could be registered at some conditions, and the distribution function is bell shaped with a maximum drifting to zero during reaction course (Fig. 4.6).

In both cases the ρ value correlates to $\langle\sigma\rangle$ and unequivocally reflects the topoisomer distribution in the reaction mixture, so FLD technique could be used for monitoring the topoisomerisation process.

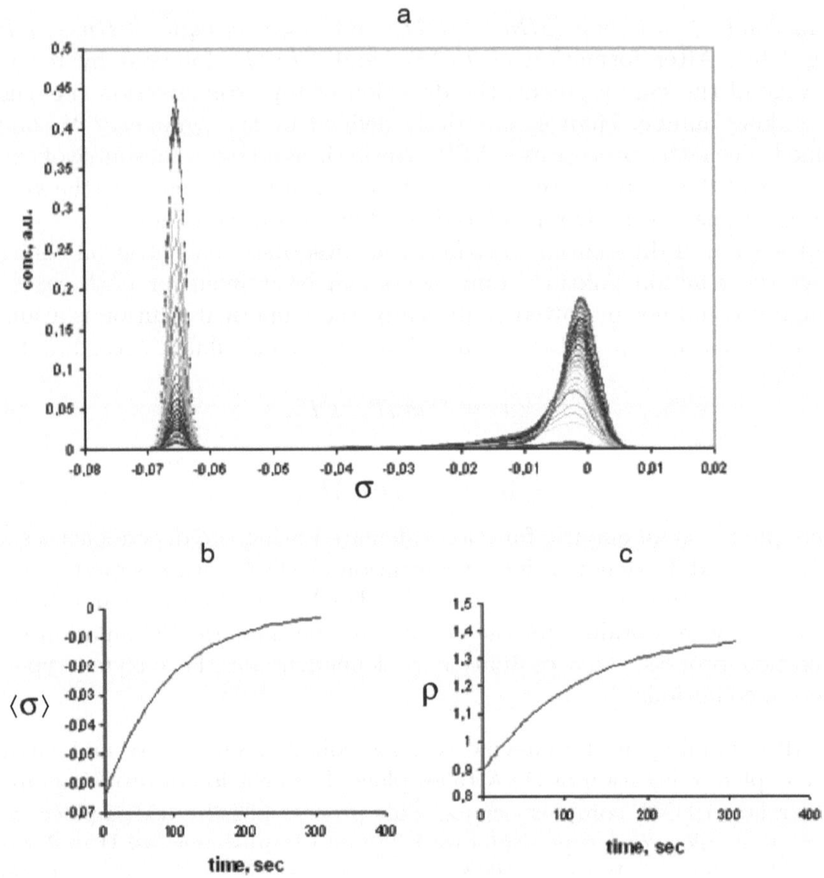

Fig. 4.5. Kinetic analysis of the scDNA transformations catalysed by topoisomerase I. (**a**) Computer simulations of the distribution of topoisomers of pTM plasmid induced by topoisomerase I. The following parameters were taken. $k_{cl} = 0.06\mathrm{s}^{-1}$, $k_l = 0.6\,\mathrm{s}^{-1}$, $k_r = 1\mathrm{s}^{-1}$, $k_d = 10^{-6}$. DNA Hooks constant $B = 1,100RT$. (**b**) $\langle\sigma\rangle$ vs time dependence calculated according to formula (4.16). (**c**) ρ vs time dependence calculated according to formula (4.17)

4.2.3 Interaction of scDNA with Eukaryotic DNA Topoisomerases

It is expedient to consider what kinds of LD signal shifts were observed in the process of pTM and BlueScript plasmid relaxation catalysed by human topoisomerase I. Some curves typical for the experiments are seen in Fig. 4.7. It is evident that the interaction between topoisomerase I and either of the plasmids results in two different pictures. In experiments with a pTM plasmid the linear dichroism of the reaction mixture rose fast. The evidence was also furnished by electrophoresis assay proofs showing that the signal shift was

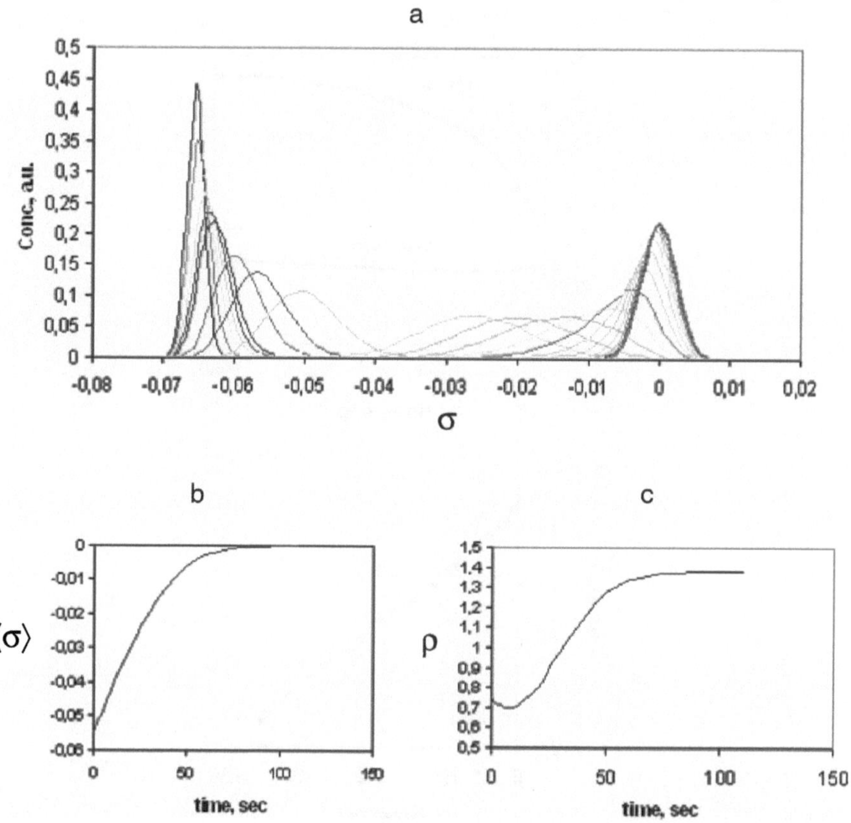

Fig. 4.6. Kinetic analysis of the scDNA transformations catalysed by topoisomerase II. (**a**) Computer simulations of the distribution of topoisomers of pTM plasmid induced by topoisomerase II. The following parameters were taken: $k_{\text{dis}} = 1\,\text{s}^{-1}$, $k_{\text{ass}} = 1{,}000\,\text{s}^{-1}$, $k_t = 5\,\text{s}^{-1}$. DNA Hooks constant $B = 1{,}150RT$. (**b**) $\langle\sigma\rangle$ vs time dependence calculated according to formula (4.16). (**c**) ρ vs time dependence calculated according to formula (4.17)

actually caused by DNA relaxation. During the reaction, the amplitude of σ shifts exactly corresponded to the difference between LD of the relaxed and supercoiled plasmids. In serial experiments (under sodium chloride concentrations exceeding 100 mM), the signal growth obeys the first-order kinetics. In contrast to the case of pTM, in the experiments with the BlueScript plasmid no change of LD signal was detected during the reaction course, although in accordance with the electrophoresis assay data the plasmid actually changes its linking number during the enzyme-mediated reaction. Actually, under any experimental conditions (e.g. under different enzyme-substrate ratio and different ionic strength), the BlueScript plasmid relaxation cannot be detected

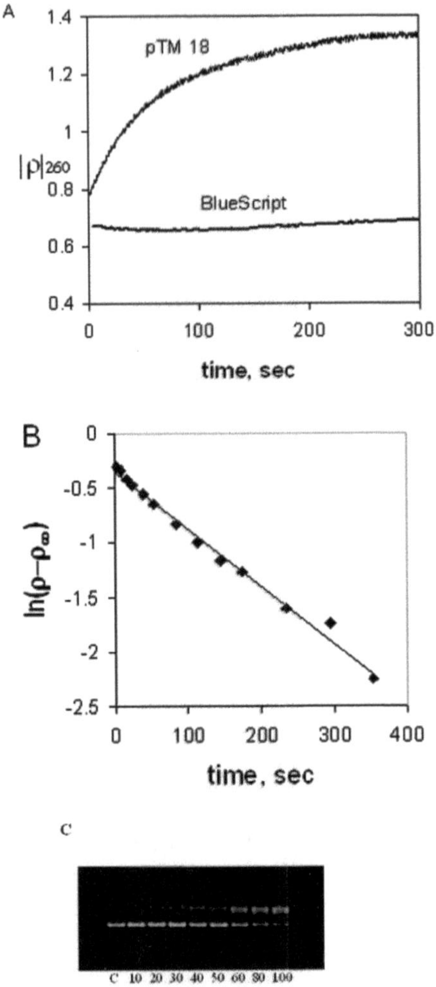

Fig. 4.7. FLD and gel electrophoresis analysis of scDNA relaxation catalysed by human topoisomerase 1–3′. (**a**) The typical FLD kinetic curves of pTM 18 and BlueScript plasmids relaxation. (**b**) The semi-logarithmic plot $\rho_t - \rho_\infty$ vs time for pTM 18 plasmid. (**c**). The agarose gel electrophoresis of the BlueScript plasmid relaxation. Experimental conditions: The FLD kinetic study of sc plasmid DNA relaxation catalysed by topoisomerase I was performed in 200 µl of the reaction mixture containing 20 mM Tris–HCl (pH 7.5), 0.5 mM EDTA, 75 µg µl15 µg $(\mu l)^{-1}$ BSA and 5–10 of sc plasmid DNA. Without special indications the experiments were performed in 100 mM NaCl. The reactions were initiated by the addition of 0.45 µg of topoisomerase I. Ten microlitres samples of the reaction mixture with BlueScript plasmid were used for electrophoresis analysis of the reaction. The enzyme kinetics was stopped by the addition of 1 µl of 10% SDS in the reaction media. Samples were analysed by 1% agarose gel electrophoresis. Gels were stained with ethidium bromide solution (1 µg ml) ml^{-1} and visualised under uv light. The experiment was performed at 25°C under the following electrophoresis conditions: TAE buffer, voltage: 5 V cm^{-1}

by FLD technique (data not shown). One can explain the latter observation taking into account that for the reaction, catalysed by topoisomerase I, calculations predict considerable amounts of highly supercoiled or relaxed DNA (and small traces of intermediate topoisomers) during entire the reaction course.

The fact that the LD changes during a topoisomerase I-catalysed reaction obeys the first-order kinetics needs an explanation. It is possible that for this reaction the K_s value does not depend on the DNA topology. To check this, we studied the relaxation of plasmid pTM 18 in the presence of sc (s) and relaxed (r) forms of BlueScript. As sc BlueScript in such experiment is virtually "invisible", both sc and relaxed BlueScript could not effect an FLD signal, so we are able to monitor the pTM 18 relaxation alone.

In both series of experiments with pTM 18, when topoisomerisation was carried out in the presence of different concentrations of relaxed or sc BlueScript, k' decreased with an increase in the mass ratio of A = [BlueScript]/[pTM 18] (Fig. 4.8). Both for relaxed and supercooled BlueScript plasmids the IC_{50} values were found to be about equal, which implies equal, affinity of the enzyme to DNA of any topology.

The insensitivity of topoisomerase I towards the DNA topology was proposed earlier based on the data of DNA relaxation kinetics in experiments with the chicken erythrocyte topoisomerase I [30]. However, latter publications contradicted this standpoint. For instance, a human topoisomerase I mutant incapable of cleaving and therefore relaxing supercoiled DNA was found to bind scDNA more than ten times more effectively as compared to the relaxed

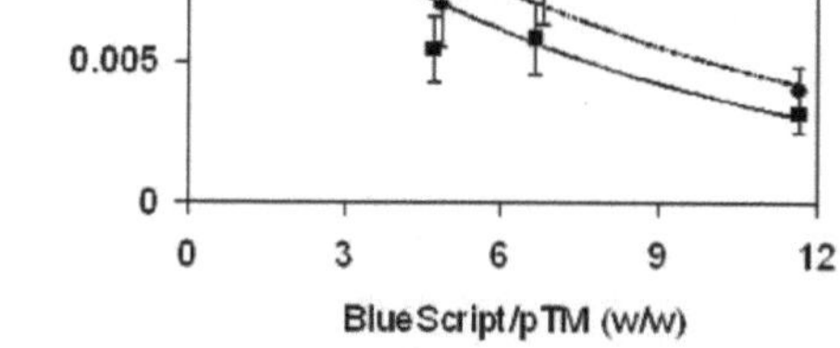

Fig. 4.8. Determination of binding parameters of topoisomerase with sc and relaxed DNA. The plasmid pTM 18 relaxation kinetics was studied in the presence of BlueScript plasmid in different concentrations taken in relaxed or sc forms. Reaction mixture contained 20 mM of Tris–HCl (pH 7.5), 0.5 mM of EDTA, 75 μg μl μl^{-1} of BSA, 5 μg of sc plasmid pTM 18 and different concentrations of relaxed or sc forms of BlueScript

form. Experiments with the native enzyme demonstrated that topoisomerase I more actively forms covalent complexes with scDNA, and more frequency produces camptothecin-induced nicks in scDNA as compared to the relaxed one [27]. These facts are considered to be sufficient proof for the suggestion that the dissociation constant of the topoisomerase I – scDNA complex is significantly lower than of the topoisomerase I-relaxed DNA one, and that catalytic activity of this enzyme is greater when DNA is highly supercoiled.

This conclusion was also supported by structural data. According to X-ray diffraction data, the contact region between the human topoisomerase I molecule and its DNA substrate is as long as one helix turn [29].

Precise calculations of the topoisomerase I – DNA contact region using Voronoi polyhedra approach using TOPOS [30] program package reveal 20 dense contact sites between protein molecule and DNA. As seen from Fig. 4.9a, b, these contact sites span more than 30 Ångstrom region on both strands of DNA helix. Therefore, the hypothesis that the enzyme could be sensitive to the tiny conformational changes affected by topological strain in DNA could not be ruled out.

At first glance, our kinetic data contradict the facts listed above. To resolve the contradiction, we made a series of calculations of the topoisomerase I reaction under the condition when the values of the dissociation constants of non-covalent enzyme substrate complex were dependent on the DNA topology state. Since the dissociation rate is fast and, probably diffusion limited, in the calculations $k_{\rm ass}$ was assumed to be topologically independent and equal to $1{,}000\ \rm min^{-1}$. The $k^n_{\rm diss}$ decreased gradually when $|\sigma|$ increases, so for $\sigma = 0.05$ the $k_{\rm diss}$ value was 10, 100, and 1,000 times lower than for the relaxed one.

Calculations made with above-mentioned sets of constants gave identical results. No difference in kinetic curve shapes was observed. Actually, it means that the process of dissociation of a topoisomerase I – highly scDNA complex had no effect on the pattern observed. The possible explanation of this statement may be presented. Topoisomerisation of Es^n into Es^{n-1} and ligation ($Es^n ES^n$) are processes in parallel. The driving force of topoisomerisation is a tension energy of scDNA and the values of $-k^n$ and k^n constants obey (4.13) and (4.14) in a broad range of n values, but the ligation rate slightly depends only on DNA topology. So, for scDNA, $k^n \gg k_1$, but for the relaxed one–the situation is quite opposite. Competition of topoisomerisation and ligation results in relaxing of the covalent enzyme–substrate complex till the degree of DNA superhelicity drops to the level close to the equilibrium. During this, the relaxing catalyst is covalently bound to DNA and is not able to dissociate. On the contrary, topoisomerase I bound to relaxed DNA exists predominantly as a non-covalent complex capable of dissociation.

In other words, there are two principally distinct mechanisms involved in the conversion of the enzyme-substrate complex, which differed in the case of scDNA and in the case of the relaxed one. For scDNA, the Briggs–Haldane mechanism works and the enzyme–substrate dissociation constant value is

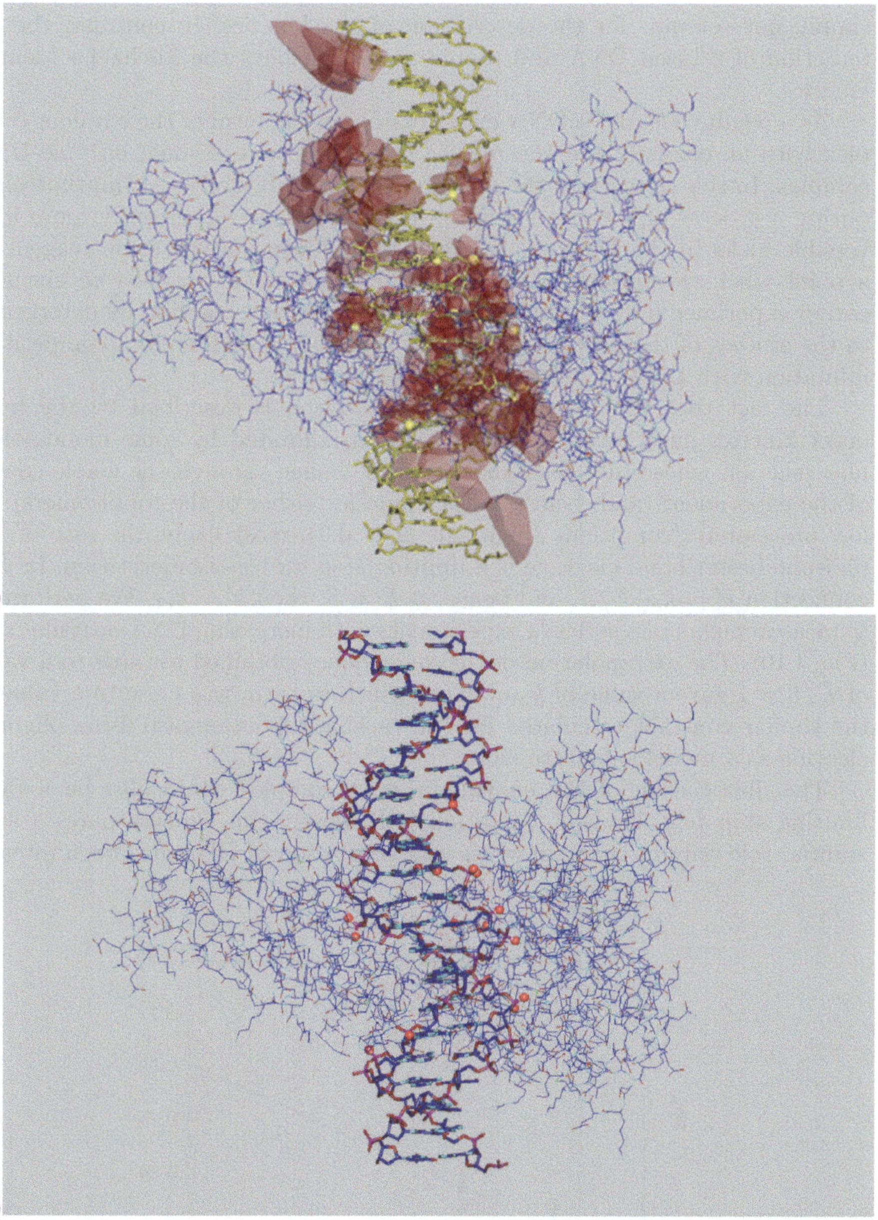

(a)

(b)

Fig. 4.9. DNA-topoisomerase interface calculated using advanced Voronoi tessellation. (**a**) Distribution of voids on the DNA-enzyme interface calculated using advanced Voronoi procedure as implemented in TOPOS [31] program package. Voids are drawn as semi-transparent polyhedra, DNA is *yellow*; (**b**) Dense DNA-protein contacts are shown as *red balls*. All DNA atoms in contact with protein are phosphate oxygen

simply not relevant for the description of the kinetics. In contrast, the interaction of relaxed DNA and topoisomerase I obeys the Michaelis–Menten kinetics.

As a result, *relaxing* scDNA exists mainly in the form of the covalent complex with an enzyme, and the *relaxed* one – in a non-covalent enzyme-DNA complex. In this connection the addition of the inhibitors, e.g. campthotecine, during a process of topoisomerisation would produce significantly more irreversible nicks in the DNA chains than the addition of the same reagent at a point when the process is over. Similar explanation can also be given to earlier experiments by stopping the topoisomerisation reaction by detergents, to the studies of the processivity of this reaction, as well as to a competitive inhibition with DNA of different topology, etc.

The fact that the topoisomerisation reaction is described by the first-order kinetics must mean that the reaction is limited by some monomolecular reaction whose rate constant does not change through the whole course of the experiment (and, hence, is the same for either of the topoisomers). In low processivity conditions (high salt, $\mu > 0.2$), most likely, the rate of the phosphodiester bond cleavage is a limiting step for topoisomerisation. In this connection $k' = k_{c1}\ E/S$, and hence, if $E = S$, then $k' = k_{c1}$. We performed k' measurements in a series of experiments with increasing DNA-enzyme ratio (Fig. 4.10). The extrapolation of the dependency obtained towards to a value of $E/S = 1$ gave a value of $k_{c1} = 0.08$, which in turn, was close to a value of the similar constant calculated for the vaccine topoisomerase using oligonucleotide as a model substrate [26].

The dissociation of the enzyme-product complex could also be a rate-limiting step (e.g. low salt conditions). However, this process plays a substantial role only in the case when a large portion of DNA has been already

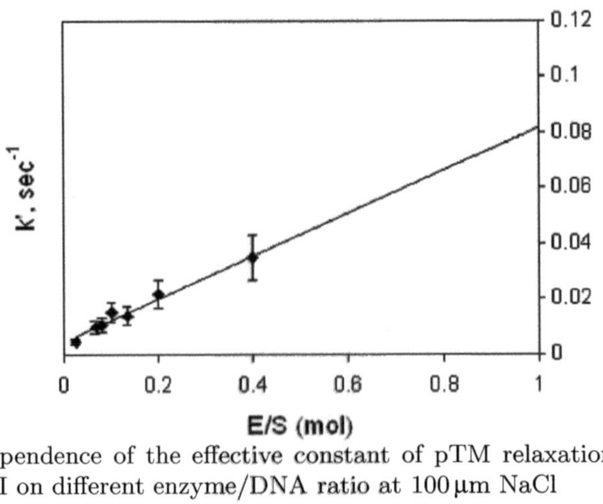

Fig. 4.10. Dependence of the effective constant of pTM relaxation catalysed by topoisomerase I on different enzyme/DNA ratio at 100 μm NaCl

relaxed. On the other hand, the mixture composition is crucial for the velocity of the process, in particular, ionic strength of the mixture. A series of experiments, in which pTM plasmid relaxation runs in media with different ionic strength has shown that the kinetic curve is split into two parts along with dropping of the ionic strength. An initial (fast) part corresponds to relaxation of the enzyme-bound DNA in a moment when the enzyme is added to the mixture. At this stage the reaction rate is only slightly dependent on the ionic strength of the buffer (Fig. 4.11a), which is actually, clear for understanding since the limiting stage at this point the shows phosphodiester bond cleavage. The second part of the curve represents a stage at which all total DNA bound with the enzyme at the fast stage, is actually fully relaxed, and the velocity of the rest of DNA is determined by a process of dissociation of enzyme from the complex with relaxed DNA (Fig. 4.11b).

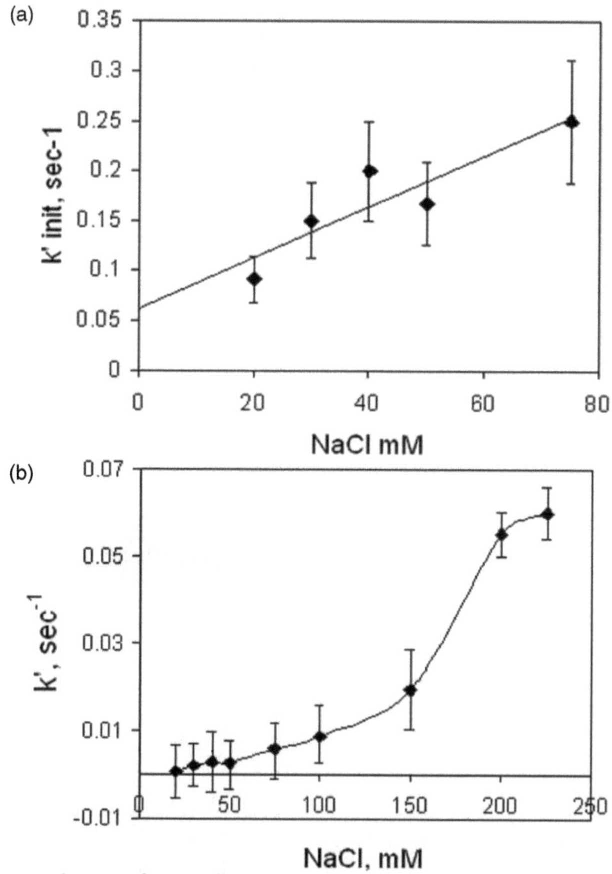

Fig. 4.11. Dependence of the effective constant of pTM relaxation catalysed by topoisomerase I. Initial (**a**) or final (**b**) part of kinetic curves on different concentrations of NaCl

Summarising these statements one can conclude that all aspects of topoisomerase I kinetics could be described within the frame of "dumb" topoisomerase acting as a reversible DNA swivel with constant friction. Our calculations reproduce all experimental kinetics curves correctly even in the case when all the kinetics constants of the equation (excluding k_b^n and k_f^n) do not depend on DNA topology.

We succeeded in a direct manner, using a non-disturbing technique to show the absence of the intermediate topoisomers in a reaction mixture during DNA topoisomerisation, catalysed by human topoisomerase I. Utilising an advanced mathematical model, we demonstrated that such topoisomers could not be detected in the mixture in the case when the reaction mechanism corresponds to Scheme 4.1. In turn, for topoisomerase II-induced relaxation of scDNA, significant amounts of intermediate topoisomers were registered (Fig. 4.12). A good coincidence takes place between experimentally observed kinetic curves and those calculated using rather simple kinetic Schemes 4.1 and 4.2. In our calculations, all kinetic constants (except two per one topoisomerisation step) were assumed to be independent of the DNA topology, so only a DNA steric strain (which obeys Hook's law) was a driving force in determining a topoisomerisation direction. So no special hypothesis describing a possible mechanism of recognition of DNA topology by topoisomerases I and II is necessary in the case of highly supercoiled DNA in contrast to those having near-to-equilibrium topology [32].

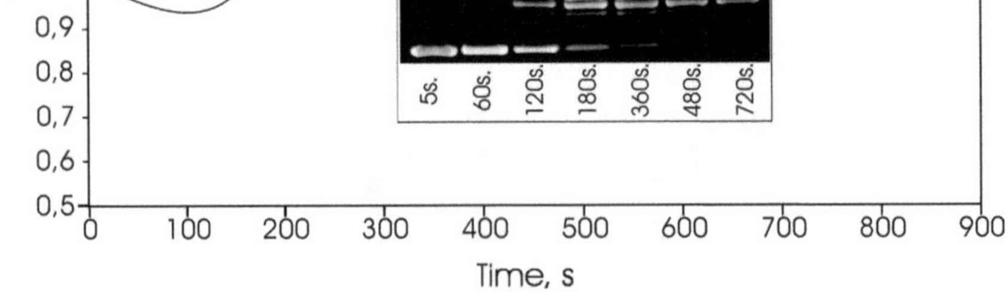

Fig. 4.12. The full-length kinetic curve of scDNA relaxation catalysed by topoisomerase II. Experimental conditions: FLD monitoring of kinetics of topoisomerisation pTM (8 μg) plasmid DNA catalysed by human topoisomerase II (4 u.) in reaction buffer (0.05 M Tris–HCl pH 8.0, 120 mM KCl, 10 mM $MgCl_2$, 0.5 mM ATP, 0.5 mM DTT, 30 μg ml^{-1} BSA). Inset: the results of electrophoretic analysis of pTM DNA in the probes of reaction mixtures corresponding to the kinetic curves taken at specified moments

4.2.4 Dynamics of Drug Targeting

The FLD technique can prove very useful in studies of the interactions of enzymes of topoisomerisation with various inhibitors and poisons [33, 34]. Such effectors are of great interest because of their high anti-tumour activity [35]. As an example, we have chosen a number of relatively well-studied compounds, some of them are already applied as anti-cancer drugs. For topoisomerase I, camptothecin (CPT) and two of its analogues, MCPT-10,11 and SN-38 (see Fig. 4.13) were taken, which are known to inhibit the process on the stage

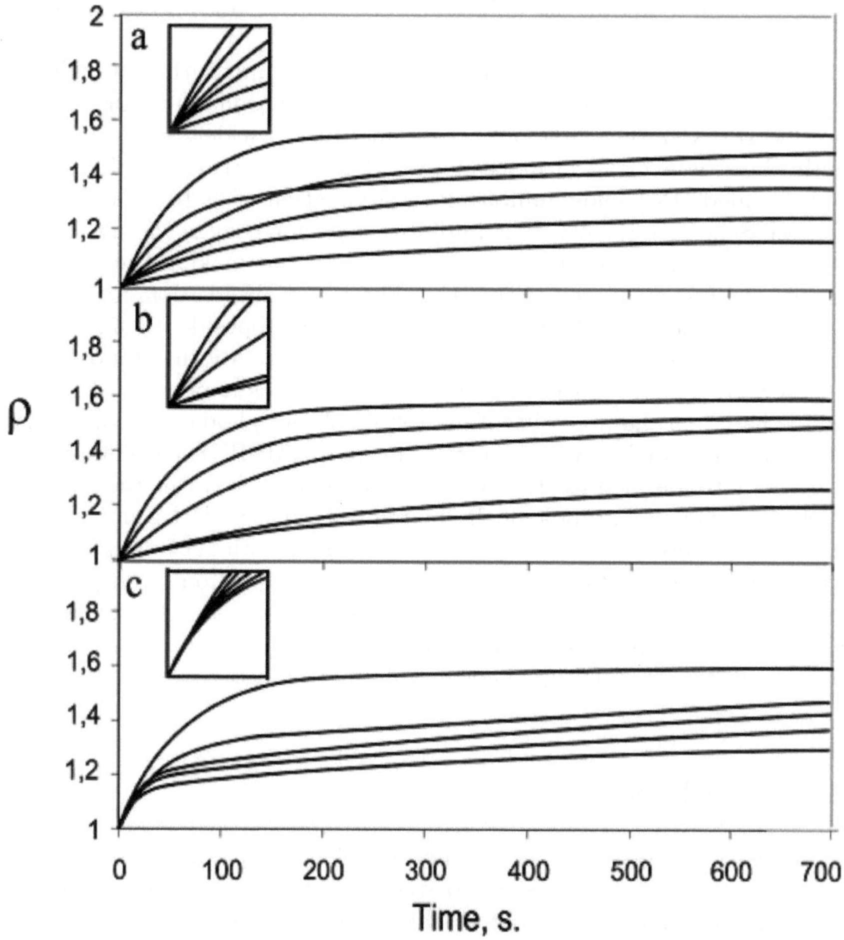

Fig. 4.13. Inhibition of topoisomerase I by specific effectors studied by FLD technique. Experimental conditions: Effect of increasing concentrations of several camptothecine analogues on the kinetics of topoisomerisation of pTM DNA (8 µg) by topoisomerase I (0.45 µg): (**a**) 0, 0.25, 0.5, 0.75, 1.25, 12.5 µM CPT; (**b**) 0, 0.05, 0.075, 0.125, 0.25 µM MCPT-10,11; (**c**) 0, 0.075, 0.15, 0.2, 0.25 µM SN-38. Insets: enlarged initial parts of the curves

of re-ligation of nicked DNA chain (see Scheme 4.1a) [36, 37]. As shown in Fig. 4.13, in the presence of the increasing concentrations of CPT and MCPT-10,11, kinetic curves retain their exponential character, while k' diminishes. I_{50} for these compounds can be estimated as 1 and 0.1 mM, respectively. On the contrary, SN-38 (Fig. 4.13) does not affect the rate of the reaction on the initial stage of the process, while effectively inhibiting it afterwards (I_{50} for the second portion is 0.2 mM). We attribute this difference to the fact that by FLD technique it becomes possible to distinguish between mechanisms of action of these closely related compounds. In our opinion, all three inhibitors hinder the re-ligation of DNA, but CPT and MCPT-10,11 impede the initial enzyme–DNA interactions as well, while SN-38 lacks this additional activity (see Scheme 4.1).

For topoisomerase II, we have chosen two compounds with totally different modes of action – etoposide [38] and adenosine-5′-phosphate-b,g-iminodiphosphate (AMPPNP). The first one, etoposide, now a widely used chemotherapeutic drug, is a classic topoisomerase II poison, acting by binding with single-stranded DNA ends, inhibiting the re-ligation of hydrolysed DNA segment (see Scheme 4.2) [39]. This inhibitor does not change the general view of kinetic curve, only slowing the whole process and making the local minimum of rotp less expressed (Fig. 4.14). The latter observation suggests that the process gets less coordinated in the presence of etoposide. Using the initial rates, we estimate I_{50} for this compound as 10 mM. Another topoisomerase II effector, AMPPNP, is the non-hydrolyzable analogue of ATP [32, 40]. Its action is based on the necessity of hydrolysis of phosphodiester bond of ATP for the completion of enzyme turnover. Thus, if AMPPNP molecule binds to one or both topoisomerase II subunits, the enzyme stays in the conformation of "closed clamp", topologically bound to the closed DNA molecule, and, therefore, is kinetically irreversibly inactivated. If both ATP and AMPPNP are present in the reaction mixture, only a certain portion of enzyme molecules is inactivated at each catalytic step. Reaction under these conditions does not proceed up to the equilibrium state, and kinetic curves reach the plateau at values of FLD signal corresponding to $s < 0$.

4.3 Conclusions

The discussed intrinsic connection between DNA topology, hydrodynamics and optical properties eneablse us to study experimentally dynamics of DNA relaxation. The fundamental functional dependency of FLD signal from the DNA topology provides us with the unique possibility of performing the continuous analysis of DNA topoisomerisation. We proved the mechanistic peculiarities of topoisomerase I, studied by traditional methods and for the first time studied the real-time kinetics for topoisomerase II. In our point

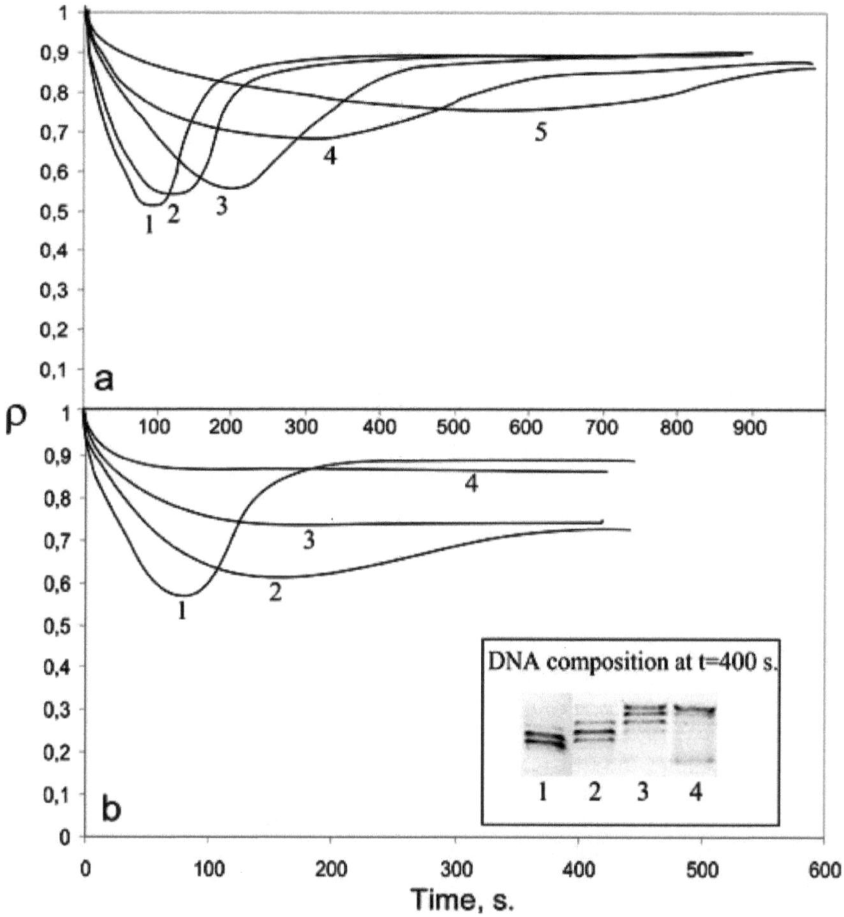

Fig. 4.14. Inhibition of topoisomerase II activity measured by FLD technique. Experimental conditions: Comparison of the kinetics of inhibition of reaction of 8 μg of pTM DNA with 4 u. of topoisomerase II by etoposide and AMPPNP. (**a**) 0 μM (curve 1), 5 μM (2), 10 μM (1), 25 μM (2), 50 μM (3) of etoposide; (**b**) ratio [AMPPNP]/([ATP]+[AMPPNP])=0 (curve 1), 0.005 (2), 0.01 (1), 0.02 (2)

of view the most prominent application of the developed approach is focused on the dynamics of drug targeting, which can open up various biomedical applications.

Acknowledgements

We would like to thank the Russian Foundation for Fundamental Investigations (00-04-48378, 05-01-00964) and Russian National Program of Support of Scientific Schools (4182.2006.1). We are very thankful for the stimulating discussions with late Professor Marat Karpeysky. We dedicate this study to his memory.

References

1. J.C. Wang, Ann. Rev. Biochem. **65**, 635–695 (1996)
2. F.H Crick, Proc. Natl Acad. Sci. USA **73**, 2639–2643 (1976)
3. F.B. Fuller, Proc. Natl Acad. Sci. USA **75**, 3557–3561 (1978)
4. A.V. Vologodskii, V.V. Anshelivich, A.V. Lukashin, M.D. Frank-Kamenetskii, Nature **280**, 294–298 (1979)
5. A.V. Vologodskii, N.R. Cozzarelli, Annu. Rev. Biophys. Biomol. Struct. **23**, 609–643 (1994)
6. J.C. Wang, L.J. Peck, K. Becherer, DNA Supercoiling and Its Effect on DNA Structure and Function. In *Cold Spring Harbor Symp. Quant. Biol.*, vol 47 (1983) pp. 85–91
7. T. Schlick, Curr. Opinion Str. Biol. **5**, 245–262 (1995)
8. J.F. Marko, E.D. Siggia, Science **265**, 506–508 (1994)
9. F.B. Fuller, Proc. Natl Acad. Sci USA **68**, 815–819 (1971)
10. J. White, Geometry and Topology of DNA and DNA-Protein Interactions, in *Proceedings of Symposia in Applied Mathematics* vol. 45 (1992), pp. 17–38
11. G. Calugareanu, Czech. Math J. **11**, 588–625 (1961)
12. M. Monastyrsky, *Topology of Gauge Fields and Condensed Matter* (Plenum Press, New York, MIR Publishers, 1993)
13. P.G. De Gennes, J. Prost, *The Physics of Liquid Crystals* (Clarendon Press, Oxford, 2nd Ed, 1993)
14. T.R. Strick, V. Croquette, D. Bensimon, Nature **404**, 901–904 (2000)
15. A.M. Shuster, G.V. Gololobov, O.A. Kvashuk et al, Science **256**, 665–667 (1992)
16. G.V. Gololobov, E.A. Chernova, D.V. Schourov et al, Proc. Natl Acad. Sci. USA **92**, 254–257 (1995)
17. A.G. Gabibov, O. Makarevitch, Methods Mol. Biol. **51**, 223–235 (1995)
18. E.A. Yakubovskaya, I.A. Kudelina, I.B. Bronshtein, A.G. Gabibov, Dokl. Akad. Nauk. **361**, 837–838 (1998)
19. P.V. Favorov, E.A. Yakubovskaya, A.V. Reshetnyak, A.G. Gabibov, Dokl. Akad. Nauk. **370**, 834–838 (2000)
20. A. Rodger, Methods Enzymol. **226**, 232–258 (1993)
21. B. Norden, M. Kubista, T. Kuruscev, Quart. Rev. Biophy. **25**, 51–170 (1992)
22. H. Yoshida, C.E. Swenberg, N.E. Geacintov, Biochemistry **26**, 1351–1358 (1987)
23. E.A. Yakubovskaya, G.V. Gololobov, I.A. Kudelina et al., Mol. Biol. (Mosk) **30**, 1378–1384 (1996)
24. V.L. Makarov, S.I. Dimitrov, Mol. Biol. (Russian) **16(3)**, 1086–1096 (1982)
25. C.E. Swenberg, S.E. Carberry, N.E. Geacintov, Biopolymers **14**, 1735–1744 (1990)
26. J.T. Stivers, S. Shuman, A.S. Mildvan, Biochemistry **33**, 327–339 (1994)
27. E.A. Yakubovskaya, A.G. Gabibov, Mol. Biol. (Mosk) **33**, 368–384 (1999)
28. J.M. Berger, Curr. Opin. Struct. Biol. **8**, 26–32 (1998)
29. L. Stewart, M.R. Redinbo, X. Oiu et al., Science **279**, 1534–1541 (1998)
30. D.E. Pulleyblank, M.J. Ellison, Biochemistry **21**, 1155–1161 (1982)

31. J. Roca, J.C. Wang, Cell **71**, 833–840 (1992)
32. A.V. Vologodskii, W. Zhang, V.V. Rybenkov et al., Proc. Natl Acad. Sci. USA **98**, 3045–3049 (2001)
33. E.A. Yakubovskaya, *Biocatalytic Transformations of Supercoiled DNA as Studied by Flow Linear Dichroism Technique* (Engelhardt Institute of Molecular Biology, Russian Academy of Sciences), Ph.D. Thesis, Moscow, (1997) pp. 49–102
34. F. Fleury, A. Sukhanova, A. Ianoul et al., J. Biol. Chem. **275**, 3501–3509 (2000)
35. T. Andoh, Biochimie **80**, 235–246 (1998)
36. E. Kjeldsen, J.Q. Svejstrup, I.I. Gromova et al., Mol. Biol. **228**, 1025–1030 (1992)
37. M. Gupta, A. Fujimori, Y. Pommier, Biochim. Biophis. Acta **1262**, 1–14 (1992)
38. K.R. Hande, Eur. J. Cancer **34**, 1514–1521 (1998)
39. S.K. Morris, J.E Lindsley, J. Biol. Chem. **274**, 30690–30696 (1999)
40. J. Roca, J.C. Wang, Cell **77**, 609–616 (1994)

5

From Tangle Fractions to DNA

L.H. Kauffman and S. Lambropoulou

Summary. This chapter draws a line from the elements of tangle fractions to the tangle model of DNA recombination. In the process, we sketch the classification of rational tangles, unoriented and oriented rational knots and the application of these subjects to DNA recombination.

5.1 Introduction

Rational knots and links are a class of alternating links of one or two unknotted components, and they are the easiest knots to make (also for Nature!). The first 25 knots, except for 8_5, are rational. Furthermore all knots and links up to ten crossings are either rational or are obtained from rational knots by insertion operations on certain simple graphs. Rational knots are also known in the literature as four-plats, Viergeflechte and twobridge knots. The lens spaces arise as twofold branched coverings along rational knots.

A rational tangle is the result of consecutive twists on neighbouring endpoints of two trivial arcs, see Definition 1. Rational knots are obtained by taking numerator closures of rational tangles (see Fig. 5.19), which form a basis for their classification. Rational knots and rational tangles are of fundamental importance in the study of DNA recombination. Rational knots and links were first considered in [1] and [2]. Treatments of various aspects of rational knots and rational tangles can be found in [3–11]. A rational tangle is associated in a canonical manner with a unique, reduced rational number or ∞, called *the fraction* of the tangle. Rational tangles are classified by their fractions by means of the following theorem:

Theorem 1. (Conway, 1970). *Two rational tangles are isotopic if and only if they have the same fraction.*

John H. Conway [4] introduced the notion of tangle and defined the fraction of a rational tangle using the continued fraction form of the tangle and the Alexander polynomial of knots. Via the Alexander polynomial, the fraction is defined for the larger class of all 2-tangles. In this study we are interested in different definitions of the fraction, and we give a self-contained exposition of the construction of the invariant fraction for arbitrary two-tangles from the bracket polynomial [12]. The tangle fraction is a key ingredient in both the classification of rational knots and in the applications of knot theory to DNA. Proofs of Theorem 1 can be found in [2], [6] p. 196 and [8, 13].

More than one rational tangle can yield the same or isotopic rational knots and the equivalence relation between the rational tangles is reflected in an arithmetic equivalence of their corresponding fractions. This is marked by a theorem due originally to Schubert [14] and reformulated by Conway [4] in terms of rational tangles.

Theorem 2. (Schubert, 1956). *Suppose that rational tangles with fractions p/q and p'/q' are given (p and q are relatively prime; similarly for p' and q'). If $K(p/q)$ and $K(p'/q')$ denote the corresponding rational knots obtained by taking numerator closures of these tangles, then $K(p/q)$ and $K(p'/q')$ are topologically equivalent if and only if*

1. *$p = p'$ and*
2. *Either $q \equiv q' (mod\, p)$ or $qq' \equiv 1 (mod\, p)$.*

This classic theorem [14] was originally proved by using an observation of Seifert that the twofold branched covering spaces of S^3 along $K(p/q)$ and $K(p'/q')$ are lens spaces, and invoking the results of Reidemeister [15] on the classification of lens spaces. Another proof using covering spaces has been given by Burde in [16]. Schubert also extended this theorem to the case of oriented rational knots and links described as two-bridge links.

Theorem 3. (Schubert, 1956). *Suppose that orientation-compatible rational tangles with fractions p/q and p'/q' are given with q and q' odd (p and q are relatively prime; similarly for p' and q') If $K(p/q)$ and $K(p'/q')$ denote the corresponding rational knots obtained by taking numerator closures of these tangles, then $K(p/q)$ and $K(p'/q')$ are topologically equivalent if and only if*

1. *$p = p'$ and*
2. *Either $q \equiv q' (mod\, 2p)$ or $qq' \equiv 1 (mod\, 2p)$.*

In [17] we give the first combinatorial proofs of Theorems 2 and 3. In this chapter we sketch the proofs in [13] and [17] of the above three theorems and we give the key examples that are behind all of our proofs. We also give some applications of Theorems 2 and 3 using our methods.

The study is organized as follows. In Sect. 5.2 we introduce two-tangles and rational tangles, Reideimeister moves, isotopies and operations. We give the definition of flyping, and state the (now-proved) Tait flyping conjecture. The Tait conjecture is used implicitly in our classification work. In Sect. 5.3 we introduce the continued fraction expression for rational tangles and its properties. We use the continued fraction expression for rational tangles to define their fractions. Then rational tangle diagrams are shown to be isotopic to alternating diagrams. The alternating form is used to obtain a canonical form for rational tangles, and we obtain a proof of Theorem 1.

Section 5.4 discusses alternate definitions of the tangle fraction. We begin with a self-contained exposition of the bracket polynomial for knots, links and tangles. Using the bracket polynomial we define a fraction $F(T)$ for arbitrary two-tangles and show that it has a list of properties that are sufficient to prove that for T rational, $F(T)$ is identical to the continued fraction value of T, as defined in Sect. 5.3. The next part of Sect. 5.4 gives a different definition of the fraction of a rational tangle, based on colouring the tangle arcs with integers. This definition is restricted to rational tangles and those tangles that are obtained from them by tangle-arithmetic operations, but it is truly elementary, depending just on a little algebra and the properties of the Reidemeister moves. Finally, we sketch yet another definition of the fraction for two-tangles that shows it to be the value of the conductance of an electrical network associated with the tangle.

Sect. 5.5 contains a description of our approach to the proof of Theorem 2, the classification of unoriented rational knots and links. The key to this approach is enumerating the different rational tangles whose numerator closure is a given unoriented rational knot or link, and confirming that the corresponding fractions of these tangles satisfy the arithmetic relations of the Theorem. Section 5.6 sketches the classification of rational knots and links that are isotopic to their mirror images. Such links are all closures of palindromic continued fraction forms of even length. Section 5.7 describes our proof of Theorem 3, the classification of oriented rational knots. The statement of Theorem 3 differs from the statement of Theorem 2 in the use of integers modulo $2p$ rather than p. We see how this difference arises in relation to matching orientations on tangles. This section also includes an explanation of the fact that fractions with even numerators correspond to rational links of two components, while fractions with odd numerators correspond to single component rational knots (the denominators are odd in both cases). Section 5.8 discusses strongly invertible rational knots and links. These correspond to palindromic continued fractions of odd length.

Section 5.9 is an introduction to the tangle model for DNA recombination. The classification of the rational knots and links, and the use of the tangle fractions is the basic topology behind the tangle model for DNA recombination. We indicate how problems in this model are reduced to properties of rational knots, links and tangles, and we show how a finite number of observations of successive DNA recombination can pinpoint the recombiation mechanism. We have included references [42–50] for the reader who is interested in delving into the background on a number of the topics that we touch in this paper.

5.2 Two-Tangles and Rational Tangles

Throughout this chapter we work with *two-tangles*. The theory of tangles was invented by John Conway [4] in his work on enumerating and classifying knots.

A two-tangle is an embedding of two arcs (homeomorphic to the interval [0,1]) and circles into a three-dimensional ball B^3 standardly embedded in Euclidean three-space S^3, such that the endpoints of the arcs go to a specific set of four points on the surface of the ball, so that the circles and the interiors of the arcs are embedded in the interior of the ball. The left-hand side of Fig. 5.1 illustrates a two-tangle. Finally, a two-tangle is *oriented* if we assign orientations to each arc and each circle. Without loss of generality, the four endpoints of a two-tangle can be arranged on a great circle on the boundary of the ball. One can then define a *diagram* of a two-tangle to be a regular projection of the tangle on the plane of this great circle. In illustrations we may replace this circle by a box.

The simplest possible two-tangles comprise two unlinked arcs, either horizontal or vertical. These are the *trivial tangles*, denoted [0] and [∞] tangles, respectively, see Fig. 5.2.

Definition 1 A two-tangle is *rational* if it can be obtained by applying a finite number of consecutive twists of neighbouring endpoints to the elementary tangles [0] or [∞].

The simplest rational tangles are the [0], the [∞], the [+1] and the [−1] tangles, as illustrated in Fig. 5.3, while the next simplest ones are:

(i) The *integer tangles*, denoted by $[n]$, made of n horizontal twists, $n \in \mathbb{Z}$.
(ii) The *vertical tangles*, denoted by $1/[n]$, made of n vertical twists, $n \in \mathbb{Z}$. These are the inverses of the integer tangles, see Fig. 5.3. This terminology will be clear soon.

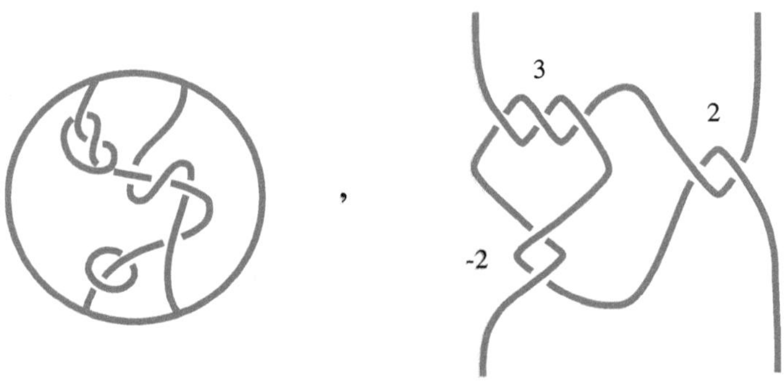

Fig. 5.1. A two-tangle and a rational tangle

Fig. 5.2. The trivial tangles [0] and [∞]

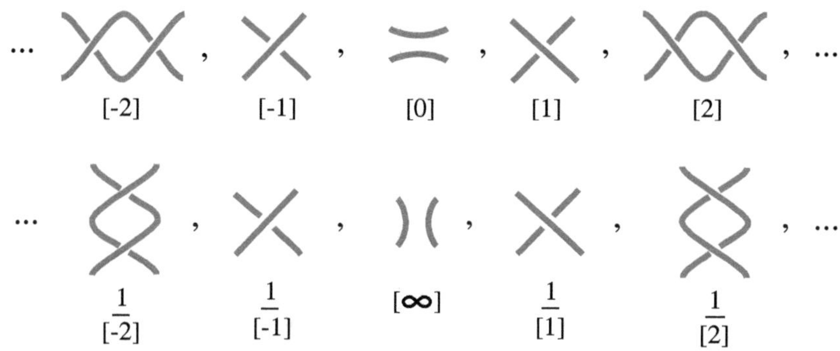

Fig. 5.3. The elementary rational tangles

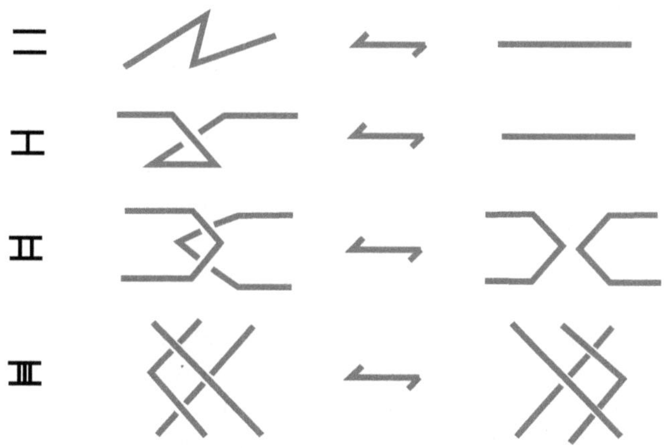

Fig. 5.4. The Reidemeister moves

Examples of rational tangles are illustrated in the right-hand side of Fig. 5.1 as well as in Figs. 5.8 and 5.17 below.

We study tangles up to *isotopy*. Two two-tangles, T, S, in B^3 are said to be *isotopic*, denoted by $T \sim S$, if they have identical configurations of their four endpoints in the boundary S^2 of the three-ball, and there is an ambient isotopy of (B^3, T) to (B^3, S) that is the identity on the boundary $(S^2, \partial T) = (S^2, \partial S)$. An ambient isotopy can be imagined as a continuous deformation of B^3 fixing the four endpoints on the boundary sphere, and bringing one tangle to the other without causing any self-intersections.

In terms of diagrams, Reidemeister [18] proved that the local moves on diagrams illustrated in Fig. 5.4 capture combinatorially the notion of ambient isotopy of knots, links and tangles in three-dimensional space. That is, if two diagrams represent knots, links or tangles that are isotopic, then the one

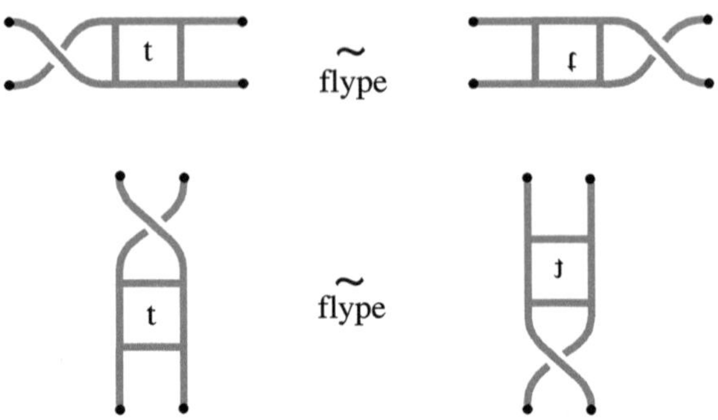

Fig. 5.5. The flype moves

diagram can be obtained from the other by a sequence of *Reidemeister moves.* In the case of tangles *the endpoints of the tangle remain fixed* and all the moves occur inside the tangle box.

Two oriented two-tangles are said to be *oriented isotopic* if there is an isotopy between them that preserves the orientations of the corresponding arcs and the corresponding circles. The diagrams of two oriented isotopic tangles differ by a sequence of *oriented Reidemeister moves*, i.e. Reidemeister moves with orientations on the little arcs that remain consistent during the moves.

From now on we will be thinking in terms of tangle diagrams. Also, we will be referring to both knots and links whenever we say "knots".

A *flype* is an isotopy move applied on a two-subtangle of a larger tangle or knot as shown in Fig. 5.5. A flype preserves the alternating structure of a diagram. Even more, flypes are the only isotopy moves needed in the statement of the celebrated Tait conjecture for alternating knots, stating that *two alternating knots are isotopic if and only if any two corresponding diagrams on S^2 are related by a finite sequence of flypes.* This was posed by P.G. Tait [19] in 1898 and proved by W. Menasco and M. Thistlethwaite, [20] in 1993.

The class of two-tangles is closed under the operations of *addition* $(+)$ and *multiplication* $(*)$ as illustrated in Fig. 5.6. Addition is accomplished by placing the tangles side-by-side and attaching the NE strand of the left tangle to the NW strand of the right tangle, while attaching the SE strand of the left tangle to the SW strand of the right tangle. The product is accomplished by placing one tangle underneath the other and attaching the upper strands of the lower tangle to the lower strands of the upper tangle.

The *mirror image* of a tangle T is denoted by $-T$ and it is obtained by switching all the crossings in T. Another operation is *rotation* accomplished by turning the tangle counter-clockwise by $90°$ in the plane. The rotation of T is denoted by T^r. The *inverse* of a tangle T, denoted by $1/T$, is defined to be $-T^r$ (See Fig. 5.6). In general, the inversion or rotation of a two-tangle is an order 4 operation. Remarkably, for rational tangles the inversion (rotation) is an order 2 operation. It is for this reason that we denote the inverse of a

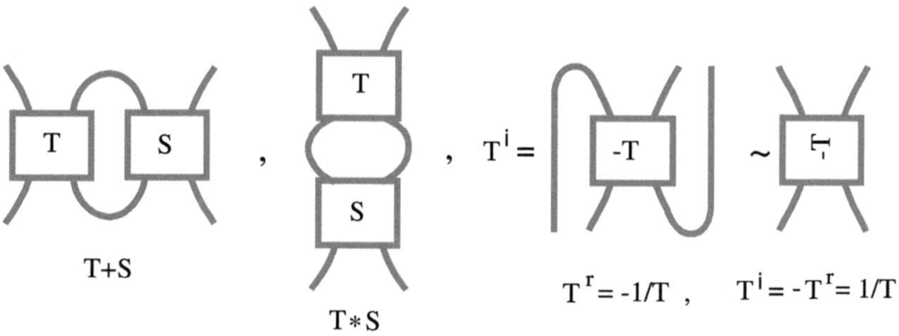

Fig. 5.6. Addition, product and inversion of two-tangles

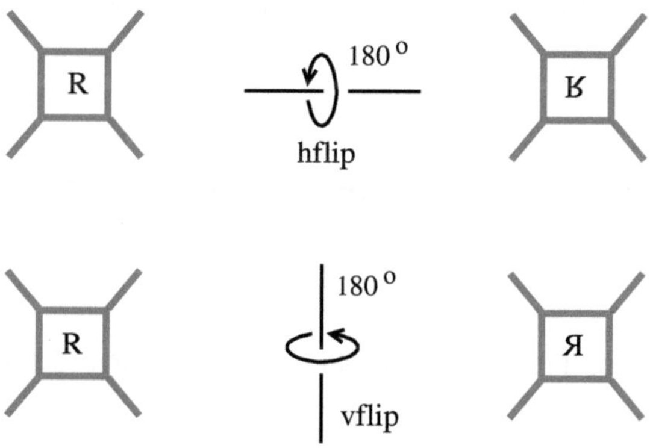

Fig. 5.7. The horizontal and the vertical flips

two-tangle T by $1/T$ or T^{-1}, and hence the rotation of the tangle T can be denoted by $-1/T = -T^{-1}$.

We now describe another operation applied on two-tangles, which turns out to be an isotopy on rational tangles. We state that R^{hflip} is the *horizontal flip* of the tangle R if R^{hflip} is obtained from R by a 180° rotation around a horizontal axis on the plane of R. Moreover, R^{vflip} is the *vertical flip* of the two-tangle R if R^{vflip} is obtained from R by a 180° rotation around a vertical axis on the plane of R (see Fig. 5.7 for illustrations). Note that a flip switches the endpoints of the tangle and, in general, a flipped tangle is not isotopic to the original one. *It is a property of rational tangles that* $T \sim T^{hflip}$ *and* $T \sim T^{vflip}$ *for any rational tangle* T. This is obvious for the tangles $[n]$ and $1/[n]$. The general proof crucially uses flypes, see [13].

The above isotopies composed consecutively yield $T \sim (T^{-1})^{-1} = (T^r)^r$ for any rational tangle T. This shows that inversion (rotation) is an operation of order 2 for rational tangles, so we can rotate the mirror image of T by 90° either counter-clockwise or clockwise to obtain T^{-1}.

Note that the twists generating the rational tangles could take place between the right, left, top or bottom endpoints of a previously created rational

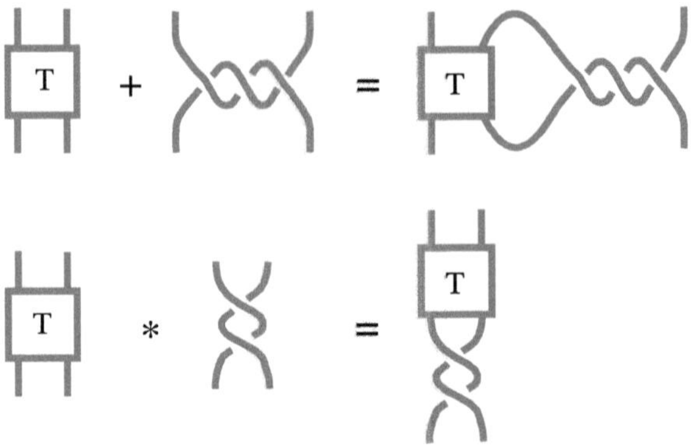

Fig. 5.8. Creating new rational tangles

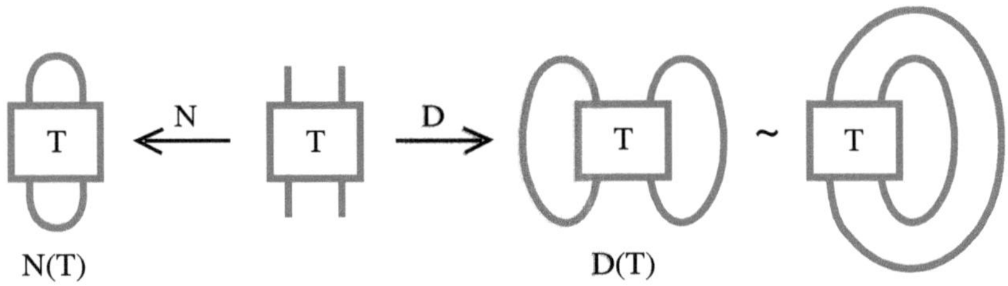

Fig. 5.9. The numerator and denominator of a two-tangle

tangle. Using flypes and flips inductively on subtangles one can always bring the twists to the right or bottom of the rational tangle. We shall then say that the rational tangle is in *standard form.* Thus a rational tangle in standard form is created by consecutive additions of the tangles $[\pm 1]$ *only on the right* and multiplications by the tangles $[\pm 1]$ *only at the bottom,* starting from the tangles $[0]$ or $[\infty]$. For example, Fig. 5.1 illustrates the tangle $(([3] * 1/[-2]) + [2])$, while Fig. 5.17 illustrates the tangle $(([3] * 1/[2]) + [2])$ in standard form. Figure 5.8 illustrates addition on the right and multiplication on the bottom by elementary tangles.

We also have the following *closing* operations, which yield two different knots: the *Numerator* of a two-tangle T, denoted by $N(T)$, obtained by joining with simple arcs the two upper endpoints and the two lower endpoints of T, and the *Denominator* of a two-tangle T, obtained by joining with simple arcs each pair of the corresponding top and bottom endpoints of T, denoted by $D(T)$ (Fig. 5.9). We have $N(T) = D(T^r)$ and $D(T) = N(T^r)$. We note that every knot or link can be regarded as the numerator closure of a two-tangle.

We obtain $D(T)$ from $N(T)$ by a $[0]$–$[\infty]$ interchange, as shown in Fig. 5.10. This "transmutation" of the numerator to the denominator is a precursor to the tangle model of a recombination event in DNA, see Sect. 5.9. The $[0]$–$[\infty]$ interchange can be described algebraically by the equations:

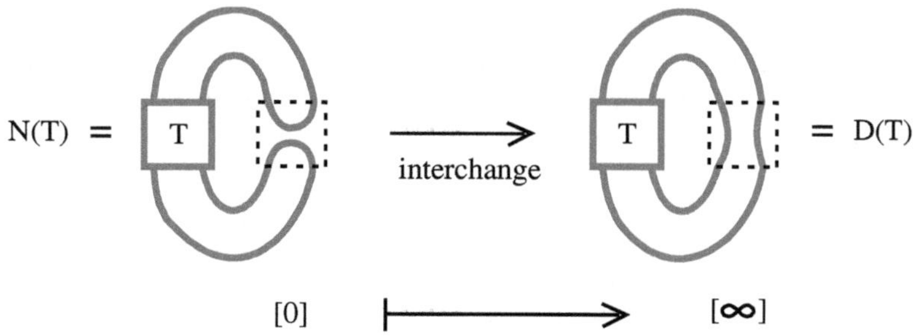

Fig. 5.10. The $[0]$–$[\infty]$ interchange

$$N(T) = N(T + [0]) \longrightarrow N(T + [\infty]) = D(T).$$

We will concentrate on the class of *rational knots and links* arising from closing the rational tangles. Even though the sum/product of rational tangles is in general not rational, the numerator (denominator) closure of the sum/product of two rational tangles is still a rational knot. It may happen that two rational tangles are not isotopic but have isotopic numerators. This is the basic idea behind the classification of rational knots, see Sect. 5.5.

5.3 Continued Fractions and the Classification of Rational Tangles

In this section we assign a fraction to a rational tangle, and we explore the analogy between rational tangles and continued fractions. This analogy culminates in a common canonical form, which is used to deduce the classification of rational tangles.

We first observe that multiplication of a rational tangle T by $1/[n]$ may be obtained as the addition of $[n]$ to the inverse $1/T$ followed by inversion. Indeed, we have:

Lemma 1. *The following tangle equation holds for any rational tangle T.*

$$T * \frac{1}{[n]} = \frac{1}{[n] + \frac{1}{T}}.$$

Thus any rational tangle can be built by a series of the following operations: Addition of $[\pm 1]$ and Inversion.

Proof. Observe that a 90° clockwise rotation of $T * 1/[n]$ produces $-[n] - 1/T$. Hence, from the above $(T * 1/[n])^r = -[n] - 1/T$, and thus $(T * 1/[n])^{-1} = [n] + 1/T$. So, taking inversions on both sides yields the tangle equation of the statement.

Definition 2 A *continued fraction in integer tangles* is an algebraic description of a rational tangle via a continued fraction built from the tangles $[a_1]$, $[a_2]$, ..., $[a_n]$ with all numerators equal to 1, namely an expression of the type:

$$[[a_1],[a_2],\ldots,[a_n]] \;:=\; [a_1] + \cfrac{1}{[a_2] + \cdots + \cfrac{1}{[a_{n-1}] + \frac{1}{[a_n]}}}$$

for $a_2,\ldots,a_n \in \mathbb{Z}-\{0\}$ and n even or odd. We allow that the term a_1 may be zero, and in this case the tangle $[0]$ may be omitted. A rational tangle described via a continued fraction in integer tangles is said to be in *continued fraction form*. The *length* of the continued fraction is arbitrary – in the previous formula illustrated with length n – whether the first summand is the tangle $[0]$ or not.

It follows from Lemma 3.2 that inductively *every rational tangle can be written in continued fraction form.* Lemma 3.2 makes it easy to write out the continued fraction form of a given rational tangle, since horizontal twists are integer additions, and multiplications by vertical twists are the reciprocals of integer additions. For example, Fig. 5.17 illustrates the rational tangle

$$[2] + \cfrac{1}{[-2] + \frac{1}{[3]}},$$

Fig. 5.17 illustrates the rational tangle

$$[2] + \cfrac{1}{[2] + \frac{1}{[3]}}.$$

Note that

$$\left([c] * \frac{1}{[b]}\right) + [a]$$

has the continued fraction form

$$[a] + \cfrac{1}{[b] + \frac{1}{[c]}} = [[a],[b],[c]].$$

For $T = [[a_1],[a_2],\ldots,[a_n]]$ the following statements are now straightforward.

1. $T + [\pm 1] = [[a_1 \pm 1],[a_2],\ldots,[a_n]]$,
2. $\frac{1}{T} = [[0],[a_1],[a_2],\ldots,[a_n]]$,
3. $-T = [[-a_1],[-a_2],\ldots,[-a_n]]$.

We now recall some facts about continued fractions. (See for example [21–24]). In this chapter we shall only consider continued fractions of the type

$$[a_1,a_2,\ldots,a_n] := a_1 + \cfrac{1}{a_2 + \cdots + \cfrac{1}{a_{n-1} + \frac{1}{a_n}}}$$

for $a_1 \in \mathbb{Z},\ a_2, \dots, a_n \in \mathbb{Z} - \{0\}$ and n even or odd. The *length* of the continued fraction is the number n whether a_1 is zero or not. Note that if for $i > 1$ all terms are positive or all terms are negative and $a_1 \neq 0$ ($a_1 = 0$) then the absolute value of the continued fraction is greater (smaller) than one. Clearly, the two simple algebraic operations *addition of* $+1$ *or* -1 and *inversion* generate inductively the whole class of continued fractions starting from zero. For any rational number p/q the following statements are straightforward.

1. There are $a_1 \in \mathbb{Z},\ a_2, \dots, a_n \in \mathbb{Z} - \{0\}$ such that $p/q = [a_1, a_2, \dots, a_n]$,
2. $p/q \pm 1 = [a_1 \pm 1, a_2, \dots, a_n]$,
3. $q/p = [0, a_1, a_2, \dots, a_n]$,
4. $-p/q = [-a_1, -a_2, \dots, -a_n]$.

We can now define the fraction of a rational tangle.

Definition 3 Let T be a rational tangle isotopic to the continued fraction form $[[a_1], [a_2], \dots, [a_n]]$. We define *the fraction* $F(T)$ *of* T to be the numerical value of the continued fraction obtained by substituting integers for the integer tangles in the expression for T, i.e.

$$F(T) := a_1 + \cfrac{1}{a_2 + \cdots + \cfrac{1}{a_{n-1} + \cfrac{1}{a_n}}} = [a_1, a_2, \dots, a_n],$$

if $T \neq [\infty]$, and $F([\infty]) := \infty = 1/0$, as a formal expression.

Remark 1 This definition is good in the sense that one can show that isotopic rational tangles always differ by flypes, and that the fraction is unchanged by flypes [13].

Clearly the tangle fraction has the following properties.

1. $F(T + [\pm 1]) = F(T) \pm 1$,
2. $F(\frac{1}{T}) = \frac{1}{F(T)}$,
3. $F(-T) = -F(T)$.

The main result about rational tangles (Theorem 1) is that two rational tangles are isotopic if and only if they have the same fraction. We will show that every rational tangle is isotopic to a unique alternating continued fraction form, and that this alternating form can be deduced from the fraction of the tangle. The theorem then follows from this observation.

Lemma 2. *Every rational tangle is isotopic to an alternating rational tangle.*

Proof. Indeed, if T has a non-alternating continued fraction form then the following configuration, shown in the left of Fig. 5.11, must occur somewhere in T, corresponding to a change of sign from one term to an adjacent term in the tangle continued fraction. This configuration is isotopic to a simpler isotopic configuration as shown in that figure.

Therefore, it follows by induction on the number of crossings in the tangle that T is isotopic to an alternating rational tangle.

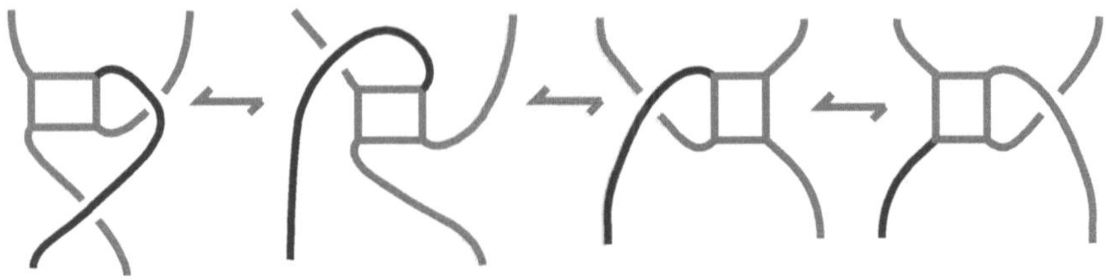

Fig. 5.11. Reducing to the alternating form

Recall that a tangle is alternating if and only if it has crossings all of the same type. Thus, *a rational tangle* $T = [[a_1], [a_2], \ldots, [a_n]]$ *is alternating if the* a_i*'s are all positive or all negative.* For example, the tangle of Fig. 5.17 is alternating.

A rational tangle $T = [[a_1], [a_2], \ldots, [a_n]]$ is said to be in *canonical form* if T is alternating and n is odd. The tangle of Fig. 5.17 is in canonical form. We note that if T is alternating and n even, then we can bring T to canonical form by breaking a_n by a unit, e.g. $[[a_1], [a_2], \ldots, [a_n]] = [[a_1], [a_2], \ldots, [a_n - 1], [1]]$, if $a_n > 0$.

The last key observation is the following well-known fact about continued fractions.

Lemma 3. *Every continued fraction* $[a_1, a_2, \ldots, a_n]$ *can be transformed to a unique canonical form* $[\beta_1, \beta_2, \ldots, \beta_m]$*, where all* β_i*'s are positive or all negative integers and* m *is odd.*

Proof. It follows immediately from Euclid's algorithm. We evaluate first $[a_1, a_2, \ldots, a_n] = p/q$, and using Euclid's algorithm we rewrite p/q in the desired form. We illustrate the proof with an example. Suppose that $p/q = 11/7$. Then

$$\frac{11}{7} = 1 + \frac{4}{7} = 1 + \frac{1}{\frac{7}{4}} = 1 + \frac{1}{1 + \frac{3}{4}} = 1 + \frac{1}{1 + \frac{1}{\frac{4}{3}}}$$

$$= 1 + \frac{1}{1 + \frac{1}{1 + \frac{1}{3}}} = [1, 1, 1, 3] = 1 + \frac{1}{1 + \frac{1}{1 + \frac{1}{2 + \frac{1}{1}}}} = [1, 1, 1, 2, 1].$$

This completes the proof.

Note that if $T = [[a_1], [a_2], \ldots, [a_n]]$ and $S = [[b_1], [b_2], \ldots, [b_m]]$ are rational tangles in canonical form with the same fraction, then it follows from this lemma that $[a_1, a_2, \ldots, a_n]$ and $[b_1, b_2, \ldots, b_m]$ are canonical continued fraction forms for the same rational number, and hence are equal term by term. Thus the uniqueness of canonical forms for continued fractions implies the uniqueness of canonical forms for rational tangles. For example, let $T = [[2], [-3], [5]]$. Then $F(T) = [2, -3, 5] = 23/14$. But $23/14 = [1, 1, 1, 1, 4]$, thus $T \sim [[1], [1], [1], [1], [4]]$, and this last tangle is the canonical form of T.

Proof (of Theorem 1). We have now assembled all the ingredients for the proof of Theorem 1. In one direction, suppose that rational tangles T and S are isotopic. Then each is isotopic to its canonical form T' and S' by a sequence of flypes. Hence the alternating tangles T' and S' are isotopic to one another. By the Tait conjecture, there is a sequence of flypes from T' to S'. Hence there is a sequence of flypes from T to S. One verifies that the fraction as we defined it is invariant under flypes. Hence T and S have the same fraction. In the other direction, suppose that T and S have the same fraction. Then, by the remark above, they have identical canonical forms to which they are isotopic, and therefore they are isotopic to each other. This completes the proof of the theorem.

5.4 Alternate Definitions of the Tangle Fraction

In the last section and in [13] the fraction of a rational tangle is defined directly from its combinatorial structure, and we verify the topological invariance of the fraction using the Tait conjecture.

In [13] we give yet another definition of the fraction for rational tangles by using colouring of the tangle arcs. There are definitions that associate a fraction $F(T)$ (including $0/1$ and $1/0$) to any two-tangle T irrespective of whether or not it is rational. The first definition is due to John Conway in [4] using the Alexander polynomial of the knots $N(T)$ and $D(T)$. In [8] an alternate definition is given that uses the bracket polynomial of the knots $N(T)$ and $D(T)$, and in [25] the fraction of a tangle is related to the conductance of an associated electrical network. In all these definitions the fraction is by definition an isotopy invariant of tangles. Below we discuss the bracket polynomial and colouring definitions of the fraction.

5.4.1 $F(T)$ Through the Bracket Polynomial

In this section we discuss the structure of the bracket state model for the Jones polynomial [12, 26] and how to construct the tangle fraction by using this technique. We first construct the bracket polynomial (state summation), which is a regular isotopy invariant (invariance under all but the Reidemeister move I). The bracket polynomial can be normalized to produce an invariant of all the Reidemeister moves. This invariant is known as the Jones polynomial [27, 28]. The Jones polynomial was originally discovered by a different method.

The *bracket polynomial*, $\langle K \rangle = \langle K \rangle(A)$, assigns to each unoriented link diagram K a Laurent polynomial in the variable A, such that

1. If K and K' are regularly isotopic diagrams, then $\langle K \rangle = \langle K' \rangle$.
2. If $K \amalg O$ denotes the disjoint union of K with an extra unknotted and unlinked component O (also called "loop" or "simple closed curve" or "Jordan curve"), then
$$\langle K \amalg O \rangle = \delta \langle K \rangle,$$

where
$$\delta = -A^2 - A^{-2}.$$

3. $\langle K \rangle$ satisfies the following formulas

$$\langle \chi \rangle = A \langle \asymp \rangle + A^{-1} \langle) (\rangle$$

$$\langle \overline{\chi} \rangle = A^{-1} \langle \asymp \rangle + A \langle) (\rangle,$$

where the small diagrams represent parts of larger diagrams that are identical except at the site indicated in the bracket. We take the convention that the letter chi, χ, denotes a crossing where *the curved line is crossing over the straight segment.* The barred letter denotes the switch of this crossing, where *the curved line is undercrossing the straight segment.* The above formulas can be summarized by the single equation

$$\langle K \rangle = A \langle S_L K \rangle + A^{-1} \langle S_R K \rangle.$$

In this text formula we have used the notations $S_L K$ and $S_R K$ to indicate the two new diagrams created by the two smoothings of a single crossing in the diagram K. That is, K, $S_L K$ and $S_R K$ differ at the site of one crossing in the diagram K. These smoothings are described as follows. Label the four regions locally incident to a crossing by the letters L and R, with L labelling the region to the left of the undercrossing arc for a traveller who approaches the overcrossing on a route along the undercrossing arc. There are two such routes, one on each side of the overcrossing line. This labels two regions with L. The remaining two are labelled R. A smoothing is of *type* L if it connects the regions labelled L, and it is of *type* R if it connects the regions labelled R, see Fig. 5.12.

It is easy to see that Properties 2 and 3 define the calculation of the bracket on arbitrary link diagrams. The choices of coefficients (A and A^{-1}) and the value of δ make the bracket invariant under the Reidemeister moves II and III (see [12]). Thus Property 1 is a consequence of the other two properties.

In order to obtain a closed formula for the bracket, we now describe it as a state summation. Let K be any unoriented link diagram. Define a *state*, S, of K to be a choice of smoothing for each crossing of K. There are two choices for smoothing a given crossing, and thus there are 2^N states of a diagram with N crossings. In a state we label each smoothing with A or A^{-1} according to the left–right convention discussed in Property 3 (see Fig. 5.12). The label is called a *vertex weight* of the state. There are two evaluations related to a state. The first one is the product of the vertex weights, denoted

$$\langle K|S \rangle.$$

The second evaluation is the number of loops in the state S, denoted

$$||S||.$$

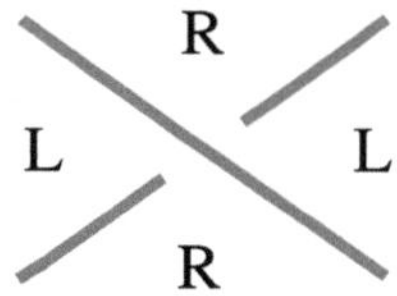

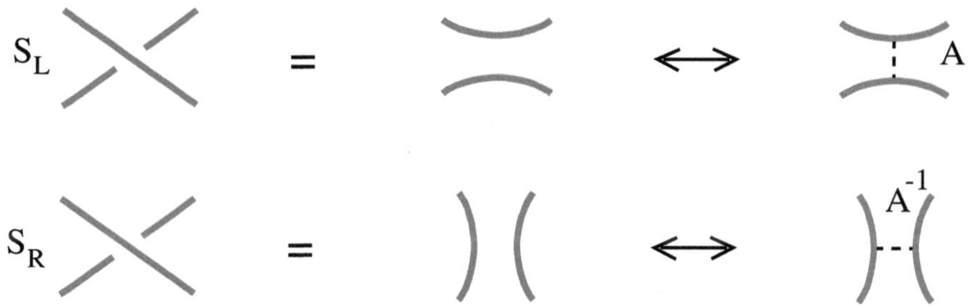

Fig. 5.12. Bracket smoothings

Define the *state summation*, $\langle K \rangle$, by the formula

$$\langle K \rangle = \sum_{S} \langle K | S \rangle \, \delta^{||S||-1}.$$

It follows from this definition that $\langle K \rangle$ satisfies the equations

$$\langle \chi \rangle = A \langle \asymp \rangle + A^{-1} \langle)(\rangle,$$

$$\langle K \amalg O \rangle = \delta \langle K \rangle,$$

$$\langle O \rangle = 1.$$

The first equation expresses the fact that the entire set of states of a given diagram is the union, with respect to a given crossing, of those states with an A-type smoothing and those with an A^{-1}-type smoothing at that crossing. The second and the third equations are clear from the formula defining the state summation. Hence this state summation produces the bracket polynomial as we have described it at the beginning of the section.

In computing the bracket, one finds the following behaviour under Reidemeister move I:

$$\langle \gamma \rangle = -A^{3} \langle \smile \rangle$$

and

$$\langle \overline{\gamma} \rangle = -A^{-3} \langle \smile \rangle,$$

where γ denotes a curl of positive type as indicated in Fig. 5.13, and $\overline{\gamma}$ indicates a curl of negative type, as also seen in this figure. The type of a curl is the sign of the crossing when we orient it locally. Our convention of signs is also given

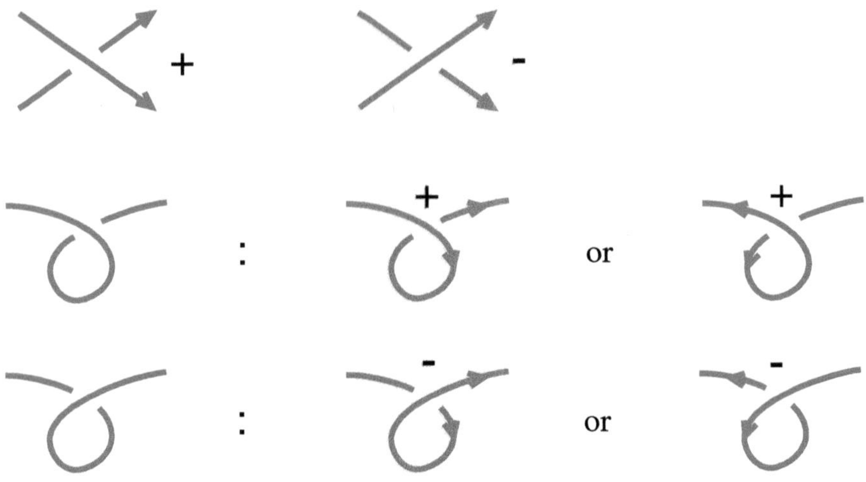

Fig. 5.13. Crossing signs and curls

in Fig. 5.13. Note that the type of a curl does not depend on the orientation we choose. The small arcs on the right-hand side of these formulas indicate the removal of the curl from the corresponding diagram.

The bracket is invariant under regular isotopy and can be normalized to an invariant of ambient isotopy by the definition

$$f_K(A) = (-A^3)^{-w(K)}\langle K\rangle(A),$$

where we chose an orientation for K, and where $w(K)$ is the sum of the crossing signs of the oriented link K. $w(K)$ is called the *writhe* of K. The convention for crossing signs is shown in Fig. 5.13.

By a change of variables one obtains the original Jones polynomial, $V_K(t)$, for oriented knots and links from the normalized bracket:

$$V_K(t) = f_K(t^{-1/4}).$$

The bracket model for the Jones polynomial is quite useful both theoretically and in terms of practical computations. One of the neatest applications is to simply compute $f_K(A)$ for the trefoil knot T and determine that $f_T(A)$ is not equal to $f_T(A^{-1}) = f_{-T}(A)$. This shows that the trefoil is not ambient isotopic to its mirror image, a fact that is quite tricky to prove by classical methods.

For two-tangles, we do smoothings on the tangle diagram until there are no crossings left. As a result, *a state of a two-tangle* consists in a collection of loops in the tangle box, plus simple arcs that connect the tangle ends. The loops evaluate to powers of δ, and what is left is either the tangle $[0]$ or the tangle $[\infty]$, since $[0]$ and $[\infty]$ are the only ways to connect the tangle inputs and outputs without introducing any crossings in the diagram. In analogy to knots and links, we can find a *state summation* formula for the *bracket of the tangle*, denoted $\langle T \rangle$, by summing over the states obtained by smoothing each crossing in the tangle. For this we define the *remainder of a state*, denoted R_S, to be either the tangle $[0]$ or the tangle $[\infty]$. Then the evaluation of $\langle T \rangle$ is given by

$$\langle T \rangle = \sum_S < T|S > \delta^{||S||} \langle R_S \rangle,$$

where $\langle T|S \rangle$ is the product of the vertex weights (A or A^{-1}) of the state S of T. The above formula is consistent with the formula for knots obtained by taking the closure $N(T)$ or $D(T)$. In fact, we have the following formula:

$$\langle N(T) \rangle = \sum_S \langle T|S \rangle \delta^{||S||} \langle N(R_S) \rangle.$$

Note that $\langle N([0]) \rangle = \delta$ and $\langle N([\infty]) \rangle = 1$. A similar formula holds for $\langle D(T) \rangle$. Thus, $\langle T \rangle$ appears as a linear combination with Laurent polynomial coefficients of $\langle [0] \rangle$ and $\langle [\infty] \rangle$, i.e. $\langle T \rangle$ takes values in the free module over $\mathbb{Z}[A, A^{-1}]$ with basis $\{\langle [0] \rangle, \langle [\infty] \rangle\}$. Notice that two elements in this module are equal iff the corresponding coefficients of the basis elements coincide. Note also that $\langle T \rangle$ is an invariant of regular isotopy with values in this module. We have just proved the following:

Lemma 4. *Let T be any two-tangle and let $\langle T \rangle$ be the formal expansion of the bracket on this tangle. Then there exist elements $n_T(A)$ and $d_T(A)$ in $\mathbb{Z}[A, A^{-1}]$, such that*

$$\langle T \rangle = d_T(A) \langle [0] \rangle + n_T(A) \langle [\infty] \rangle,$$

and $n_T(A)$ and $d_T(A)$ are regular isotopy invariants of the tangle T.

In order to evaluate $\langle N(T) \rangle$ in the formula above we need only apply the closure N to $[0]$ and $[\infty]$. More precisely, we have:

Lemma 5. $\langle N(T) \rangle = d_T \delta + n_T$ *and* $\langle D(T) \rangle = d_T + n_T \delta$.

Proof. Since the smoothings of crossings do not interfere with the closure (N or D), the closure will carry through linearly to the whole sum of $\langle T \rangle$. Thus,

$$\langle N(T) \rangle = d_T(A) \langle N([0]) \rangle + n_T(A) \langle N([\infty]) \rangle = d_T(A)\delta + n_T(A),$$

$$\langle D(T) \rangle = d_T(A) \langle D([0]) \rangle + n_T(A) \langle D([\infty]) \rangle = d_T(A) + n_T(A)\delta.$$

We define now the *polynomial fraction*, $frac_T(A)$, of the two-tangle T to be the ratio

$$frac_T(A) = \frac{n_T(A)}{d_T(A)}$$

in the ring of fractions of $\mathbb{Z}[A, A^{-1}]$ with a formal symbol ∞ adjoined.

Lemma 6. *$frac_T(A)$ is an invariant of ambient isotopy for two-tangles.*

Proof. Since d_T and n_T are regular isotopy invariants of T, it follows that $frac_T(A)$ is also a regular isotopy invariant of T. Suppose now $T\gamma$ is T with a curl added. Then $\langle T\gamma\rangle = (-A^3)\langle T\rangle$ (same remark for $\bar{\gamma}$). So, $n_{T\gamma}(A) = -A^3 n_T(A)$ and $d_{T\gamma}(A) = -A^3 d_T(A)$. Thus, $n_{T\gamma}/d_{T\gamma} = n_T/d_T$. This shows that $frac_T$ is also invariant under the Reidemeister move I, and hence an ambient isotopy invariant.

Lemma 7. *Let T and S be two two-tangles. Then, we have the following formula for the bracket of the sum of the tangles.*

$$\langle T+S\rangle = d_T d_S\langle[0]\rangle + (d_T n_S + n_T d_S + n_S\delta)\langle[\infty]\rangle.$$

Thus

$$frac_{T+S} = frac_T + frac_S + \frac{n_S\delta}{d_T d_S}.$$

Proof. We do first the smoothings in T leaving S intact, and then in S:

$$\begin{aligned}
<T+S> &= d_T\langle[0]+S\rangle + n_T\langle[\infty]+S\rangle \\
&= d_T\langle S\rangle + n_T\langle[\infty]+S\rangle \\
&= d_T(d_S\langle[0]\rangle + n_S\langle[\infty]\rangle) \\
&\quad + n_T(d_S\langle[\infty]+[0]\rangle + n_S\langle[\infty]+[\infty]\rangle) \\
&= d_T(d_S\langle[0]\rangle + n_S\langle[\infty]\rangle) + n_T(d_S\langle[\infty]\rangle + n_S\delta\langle[\infty]\rangle) \\
&= d_T d_S\langle[0]\rangle + (d_T n_S + n_T d_S + n_S\delta)\langle[\infty]\rangle.
\end{aligned}$$

Thus, $n_{T+S} = (d_T n_S + n_T d_S + n_S\delta)$ and $d_{T+S} = d_T d_S$. A straightforward calculation gives now $frac_{T+S}$.

As we see from Lemma 4, $frac_T(A)$ will be additive on tangles if

$$\delta = -A^2 - A^{-2} = 0.$$

Moreover, from Lemma 2 we have for $\delta = 0$, $\langle N(T)\rangle = n_T$, $\langle D(T)\rangle = d_T$. This nice situation will be our main object of study in the rest of this section. Now, if we set $A = \sqrt{i}$ where $i^2 = -1$, then it is

$$\delta = -A^2 - A^{-2} = -i - i^{-1} = -i + i = 0.$$

For this reason, we shall henceforth assume that A takes the value $\sqrt{i}$. So $\langle K\rangle$ will denote $\langle K\rangle(\sqrt{i})$ for any knot or link K.

We now define the *two-tangle fraction* $F(T)$ by the following formula:

$$F(T) = i\,\frac{n_T(\sqrt{i})}{d_T(\sqrt{i})}.$$

We will let $n(T) = n_T(\sqrt{i})$ and $d(T) = d_T(\sqrt{i})$, so that

$$F(T) = i\,\frac{n(T)}{d(T)}.$$

Lemma 8. *The two-tangle fraction has the following properties.*

1. $F(T) = i\,\langle N(T)\rangle / \langle D(T)\rangle$, *and it is a real number or* ∞,
2. $F(T+S) = F(T) + F(S)$,
3. $F([0]) = \frac{0}{1}$,
4. $F([1]) = \frac{1}{1}$,
5. $F([\infty]) = \frac{1}{0}$,
6. $F(-T) = -F(T)$, *in particular* $F([-1]) = -\frac{1}{1}$,
7. $F(1/T) = 1/F(T)$,
8. $F(T^r) = -1/F(T)$.

As a result we conclude that for a tangle obtained by arithmetic operations from integer tangles $[n]$, the fraction of that tangle is the same as the fraction obtained by doing the same operations to the corresponding integers. (This will be studied in detail in the next section.)

Proof. The formula $F(T) = i\,\langle N(T)\rangle / \langle D(T)\rangle$ and Statement 2 follow from the observations above about $\delta = 0$. In order to show that $F(T)$ is a real number or ∞ we first consider $\langle K\rangle := \langle K\rangle(\sqrt{i})$, for K a knot or link, as in the hypotheses prior to the lemma. Then we apply this information to the ratio $i\,\langle N(T)\rangle / \langle D(T)\rangle$.

Let K be any knot or link. We claim that then $\langle K\rangle = \omega p$, where ω is a power of $\sqrt{i}$ and p is an integer. In fact, we will show that each non-trivial state of K contributes $\pm\omega$ to $\langle K\rangle$. In order to show this, we examine how to get from one non-trivial state to another. It is a fact that, for any two states, we can get from one to the other by resmoothing a subset of crossings. It is possible to get from any single loop state (and only single loop states of K contribute to $\langle K\rangle$, since $\delta = 0$) to any other single loop state by a series of *double resmoothings.* In a double resmoothing we resmooth two crossings, such that one of the resmoothings disconnects the state and the other reconnects it. (See Fig. 5.14 for an illustration.) Now consider the effect of a double resmoothing on the evaluation of one state. Two crossings change. If one is labelled A and the other A^{-1}, then there is no net change in the evaluation of the state. If both are A, then we go from A^2P (P is the rest of the product of state labels) to $A^{-2}P$. But $A^2 = i$ and $A^{-2} = -i$. Thus if one state contributes $\omega = ip$, then the other state contributes $-\omega = -ip$. These remarks prove the claim.

Now, a state that contributes non-trivially to $N(T)$ must have the form of the tangle $[\infty]$. We will show that if S is a state of T contributing non-trivially to $\langle N(T)\rangle$ and S' a state of T contributing non-trivially to $\langle D(T)\rangle$, then $\langle S\rangle / \langle S'\rangle = \pm i$. Here $\langle S\rangle$ denotes the product of the vertex weights for S, and $< S' >$ is the product of the vertex weights for S'. If this ratio is verified

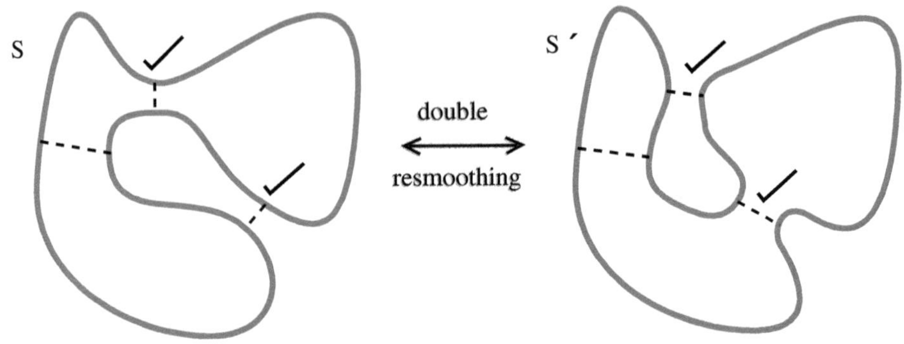

Fig. 5.14. A double resmoothing

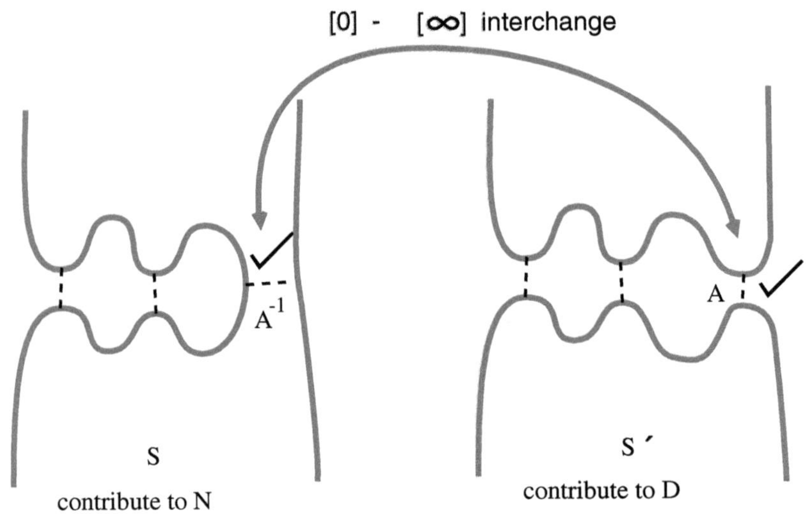

Fig. 5.15. Non-trivial states

for some pair of states S, S', then it follows from the first claim that it is true for all pairs of states, and that $\langle N(T)\rangle = \omega p$, $\langle D(T)\rangle = \omega' q$, $p, q \in \mathbb{Z}$ and $\omega/\omega' = \langle S\rangle/\langle S'\rangle = \pm i$. Hence $\langle N(T)\rangle/\langle D(T)\rangle = \pm i\, p/q$, where p/q is a rational number (or $q = 0$). This will complete the proof that $F(T)$ is real or ∞.

To see this second claim we consider specific pairs of states as in Fig. 5.15. We have illustrated representative states S and S' of the tangle T. We obtain S' from S by resmoothing at one site that changes S from an $[\infty]$ tangle to the $[0]$ tangle underlying S'. Then $\langle S\rangle/\langle S'\rangle = A^{\pm 2} = \pm i$. If there is no such resmoothing site available, then it follows that $D(T)$ is a disjoint union of two diagrams, and hence $\langle D(T)\rangle = 0$ and $F(T) = \infty$. This does complete the proof of Statement 1.

At $\delta = 0$ we also have:

$\langle N([0]\rangle = 0$, $\langle D([0])\rangle = 1$, $\langle N([\infty])\rangle = 1$, $\langle D([\infty])\rangle = 0$, and so, the evaluations 3–5 are easy. For example, note that

$$\langle [1]\rangle = A\langle [0]\rangle + A^{-1}\langle [\infty]\rangle,$$

hence

$$F([1]) = i\,\frac{A^{-1}}{A} = i\,A^{-2} = i\,(i^{-1}) = 1.$$

To have the fraction value 1 for the tangle [1] is the reason that in the definition of $F(T)$ we normalized by i. Statement 6 follows from the fact that the bracket of the mirror image of a knot K is the same as the bracket of K, but with A and A^{-1} switched. For proving 7 we observe first that for any 2-tangle T, $d(1/T) = \overline{n(T)}$ and $n(1/T) = \overline{d(T)}$, where the overline denotes the complex conjugate. Complex conjugates occur because $A^{-1} = \overline{A}$ when $A = \sqrt{i}$. Now, since $F(T)$ is real, we have

$F(\frac{1}{T}) = i\,\overline{d(T)}/\overline{n(T)} = \overline{-i\,d(T)/n(T)} = \overline{1/(i\,n(T)/d(T))} = \overline{1/F(T)} = 1/F(T)$.

Statement 8 follows immediately from 6 and 7. This completes the proof.

For a related approach to the well definedness of the two-tangle fraction, the reader should consult [29]. The double resmoothing idea originates from [30].

Remark 2 For any knot or link K we define the *determinant* of K by the formula

$$Det(K) := |\langle K\rangle(\sqrt{i})|,$$

where $|z|$ denotes the modulus of the complex number z. Thus we have the formula

$$|F(T)| = \frac{Det(N(T))}{Det(D(T))}$$

for any two-tangle T.

In other approaches to the theory of knots, the determinant of the knot is actually the determinant of a certain matrix associated either to the diagram for the knot or to a surface whose boundary is the knot. (See [7, 24] for more information on these connections). Conway's original definition of the fraction [4] is $\Delta_{N(T)}(-1)/\Delta_{D(T)}(-1)$ where $\Delta_K(-1)$ denotes the evaluation of the Alexander polynomial of a knot K at the value -1. In fact, $|\Delta_K(-1)| = Det(K)$, and with appropriate attention to signs, the Conway definition and our definition using the bracket polynomial coincide for all two-tangles.

5.4.2 The Fraction Through Colouring

We conclude this section by giving an alternate definition of the fraction that uses the concept of colouring of knots and tangles. We colour the arcs of the knot/tangle with integers, using the basic colouring rule that if two undercrossing arcs coloured α and γ meet at an overcrossing arc coloured β, then $\alpha + \gamma = 2\beta$. We often think of one of the undercrossing arc colours as determined by the other two colours. Then one writes $\gamma = 2\beta - \alpha$.

It is easy to verify that this colouring method is invariant under the Reidemeister moves in the following sense: Given a choice of colouring for the tangle/knot, there is a way to re-colour it each time a Reidemeister move is performed, so that no change occurs to the colours on the external strands of the tangle (so that we still have a valid colouring). This means that a colouring potentially contains topological information about a knot or a tangle. In colouring a knot (and also many non-rational tangles) it is usually necessary to restrict the colours to the set of integers modulo N for some modulus N. For example, in Fig. 5.16 it is clear that the colour set $\mathbb{Z}/3\mathbb{Z} = \{0, 1, 2\}$ is forced for colouring a trefoil knot. When there exists a colouring of a tangle by integers, so that it is not necessary to reduce the colours over some modulus we shall say that the tangle is *integrally colourable.*

It turns out that *every rational tangle is integrally colourable:* To see this choose two "colours" for the initial strands (e.g. the colours 0 and 1) and colour the rational tangle as you create it by successive twisting. We call the colours on the initial strands the *starting colours.* (see Fig. 5.17 for an example). It is important that we start colouring from the initial strands, because then the colouring propagates automatically and uniquely. If one starts from somewhere else, one might get into an edge with an undetermined colour. The resulting coloured tangle now has colours assigned to its external strands at the northwest, northeast, southwest and southeast positions. Let $NW(T)$, $NE(T)$, $SW(T)$ and $SE(T)$ denote these respective colours of the coloured tangle T and define the *colour matrix of* T, $M(T)$, by the equation

$$M(T) = \begin{bmatrix} NW(T) & NE(T) \\ SW(T) & SE(T) \end{bmatrix}.$$

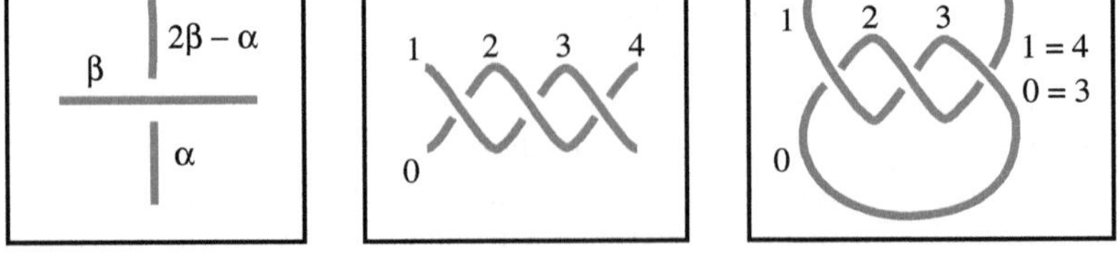

Fig. 5.16. The colouring rule, integral and modular colouring

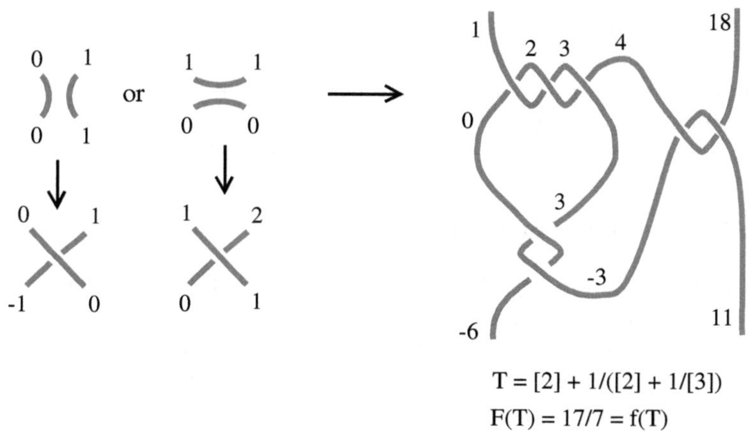

Fig. 5.17. Colouring rational tangles

Definition 4 To a rational tangle T with colour matrix $M(T) = \begin{bmatrix} a & b \\ c & d \end{bmatrix}$ we associate the number

$$f(T) := \frac{b-a}{b-d} \in \mathbb{Q} \cup \infty.$$

It turns out that the entries a, b, c, d of a colour matrix of a rational tangle satisfy the "diagonal sum rule": $a + d = b + c$.

Proposition 1 *The number $f(T)$ is a topological invariant associated with the tangle T. In fact, $f(T)$ has the following properties:*

1. $f(T + [\pm 1]) = f(T) \pm 1,$
2. $f(-\frac{1}{T}) = -\frac{1}{f(T)},$
3. $f(-T) = -f(T),$
4. $f(\frac{1}{T}) = \frac{1}{f(T)},$
5. $f(T) = F(T).$

Thus the colouring fraction is identical to the arithmetical fraction defined earlier.

It is easy to see that $f([0]) = \frac{0}{1}$, $f([\infty]) = \frac{1}{0}$, $f([\pm 1]) = \pm 1$. Hence Statement 5 follows by induction. For proofs of all statements above as well as for a more general set-up we refer the reader to our study [13]. This definition is quite elementary, but applies only to rational tangles and tangles generated from them by the algebraic operations of "+" and "$*$".

In Fig. 5.17 we have illustrated a colouring over the integers for the tangle $[[2], [2], [3]]$ such that every edge is labelled by a different integer. This is always the case for an alternating rational tangle diagram T. For the numerator closure $N(T)$ one obtains a colouring in a modular number system. For example in Fig. 5.17 the colouring of $N(T)$ will be in $\mathbb{Z}/17\,\mathbb{Z}$, and it is easy to check that the labels remain distinct in this example. For rational tangles, this is always the case when $N(T)$ has a prime determinant, see [13] and [31]. It is part of a more general conjecture about alternating knots and links [32, 33].

5.4.3 The Fraction Through Conductance

Conductance is a quantity defined in electrical networks as the inverse of resistance. For pure resistances, conductance is a positive quantity. Negative conductance corresponds to amplification, and is commonly included in the physical formalism. One defines the conductance between two vertices in a graph (with positive or negative conductance weights on the edges of the graph) as a sum of weighted trees in the graph divided by a sum of weighted trees of the same graph, but with the two vertices identified. This definition allows negative values for conductance and it agrees with the classical one. Conductance satisfies familiar laws of parallel and series connection as well as a star-triangle relation.

By associating to a given knot or link diagram the corresponding signed checkerboard graph (see [13, 25] for a definition of this well-known association of graph to link diagram), one can define [25] the conductance of a knot or link between any two regions that receive the same colour in the checkerboard graph. The conductance of the link between these two regions is an isotopy invariant of the link (with motion restricted to Reidemeister moves that do not pass across the selected regions). This invariance follows from properties of series/parallel connection and the star-triangle relation. These circuit laws turn out to be images of the Reidemeister moves under the translation from knot or link diagram to checkerboard graph! For a two-tangle we take the conductance to be the conductance of the numerator of the tangle, between the two bounded regions adjacent to the closures at the top and bottom of the tangle.

The conductance of a two-tangle turns out to be the same as the fraction of the tangle. This provides yet another way to define and verify the isotopy invariance of the tangle fraction for any two-tangle.

5.5 The Classification of Unoriented Rational Knots

By taking their numerators or denominators rational tangles give rise to a special class of knots, *the rational knots*. We have seen so far that rational tangles are directly related to finite continued fractions. We carry this insight further into the classification of rational knots (Schubert's theorems). In this section we consider unoriented knots, and by Remark 3.1 we will be using the three-strand-braid representation for rational tangles with odd number of terms. Also, by Lemma 2, we may assume all rational knots to be alternating.

Note that we only need to take numerator closures, since the denominator closure of a tangle is simply the numerator closure of its rotate.

As already mentioned in the introduction, it may happen that two rational tangles are non-isotopic but have isotopic numerators. The simplest instance of this phenomenon is adding n twists at the bottom of a tangle T, see Fig. 5.18. This operation does not change the knot $N(T)$, i.e. $N(T*1/[n]) \sim N(T)$, but it does change the tangle, since $F(T*1/[n]) = F(1/([n]+1/T)) = 1/(n + 1/F(T))$; so, if $F(T) = p/q$, then $F(T*1/[n]) = p/(np+q)$. Hence, if we set $np + q = q'$ we have $q \equiv q' (\operatorname{mod} p)$, just as Theorem 2 dictates. Note that reducing all possible bottom twists implies $|p| > |q|$.

Another key example of the arithmetic relationship of the classification of rational knots is illustrated in Fig. 5.19. Here we see that the "palindromic" tangles

$$T = [[2],[3],[4]] = [2] + \cfrac{1}{[3] + \frac{1}{[4]}}$$

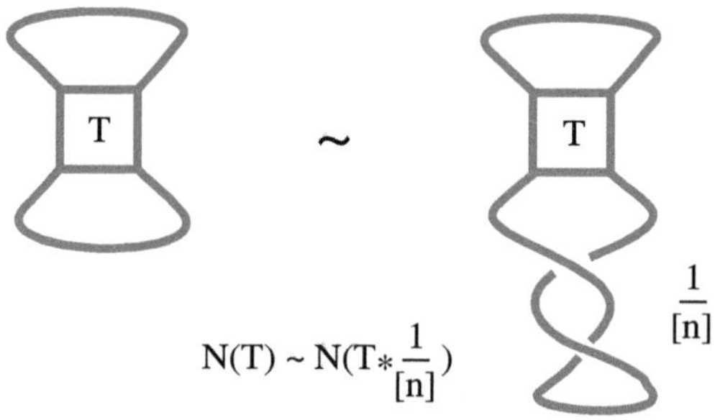

Fig. 5.18. Twisting the bottom of a tangle

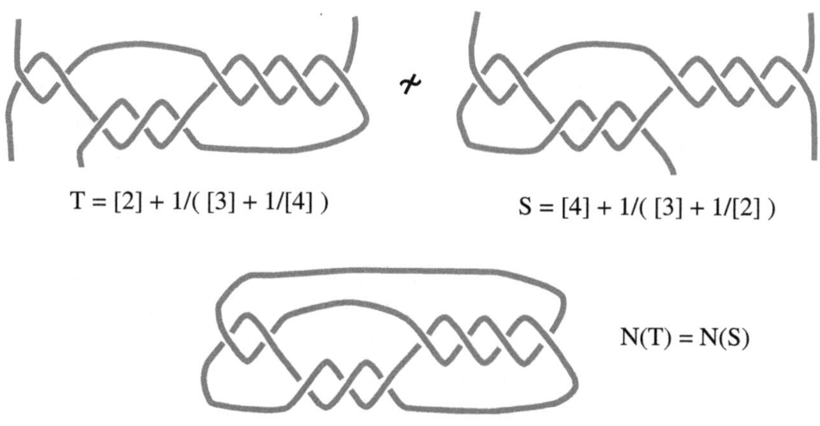

Fig. 5.19. An instance of the palindrome equivalence

and

$$S = [[4], [3], [2]] = [4] + \frac{1}{[3] + \frac{1}{[2]}}$$

both close to the same rational knot, shown at the bottom of the figure. The two tangles are different, since they have different corresponding fractions:

$$F(T) = 2 + \frac{1}{3 + \frac{1}{4}} = \frac{30}{13} \quad \text{and} \quad F(S) = 4 + \frac{1}{3 + \frac{1}{2}} = \frac{30}{7}.$$

Note that the product of 7 and 13 is congruent to 1 modulo 30.

More generally, consider the following two fractions:

$$F = [a, b, c] = a + \frac{1}{b + \frac{1}{c}} \quad \text{and} \quad G = [c, b, a] = c + \frac{1}{b + \frac{1}{a}}.$$

We find that

$$F = a + c\,\frac{1}{cb + 1} = \frac{abc + a + c}{bc + 1} = \frac{P}{Q},$$

while

$$G = c + a\,\frac{1}{ab + 1} = \frac{abc + c + a}{ab + 1} = \frac{P}{Q'}.$$

Thus we found that $F = P/Q$ and $G = P/Q'$, where

$$QQ' = (bc + 1)(ab + 1) = ab^2c + ab + bc + 1 = bP + 1.$$

Assuming that a, b and c are integers, we conclude that

$$QQ' \equiv 1\,(\mathrm{mod}\,P).$$

This pattern generalizes to arbitrary continued fractions and their palindromes (obtained by reversing the order of the terms), i.e. *If* $\{a_1, a_2, \ldots, a_n\}$ *is a collection of* n *non-zero integers, and if* $A = [a_1, a_2, \ldots, a_n] = P/Q$ *and* $B = [a_n, a_{n-1}, \ldots, a_1] = P'/Q'$, *then* $P = P'$ *and*$QQ' \equiv (-1)^{n+1}(\mathrm{mod}\,P)$. We will be referring to this as the palindrome theorem. The palindrome theorem is a known result about continued fractions. (For example, see [5] and [17]). Note that we need n to be odd in the previous congruence. This agrees with Remark 3.1 that without loss of generality the terms in the continued fraction of a rational tangle may be assumed to be odd.

Finally, Fig. 5.20 illustrates another basic example for the unoriented Schubert theorem. The two tangles $R = [1]+1/[2]$ and $S = [-3]$ are non-isotopic by the Conway theorem, since $F(R) = 1 + 1/2 = 3/2$ while $F(S) = -3 = 3/-1$. But they have isotopic numerators: $N(R) \sim N(S)$, the left-handed trefoil. Now 2 is congruent to -1 modulo 3, confirming Theorem 2.

We now analyse the above example in general. From the analysis of the bottom twists we can assume without loss of generality that a rational tangle R

Fig. 5.20. An example of the special cut

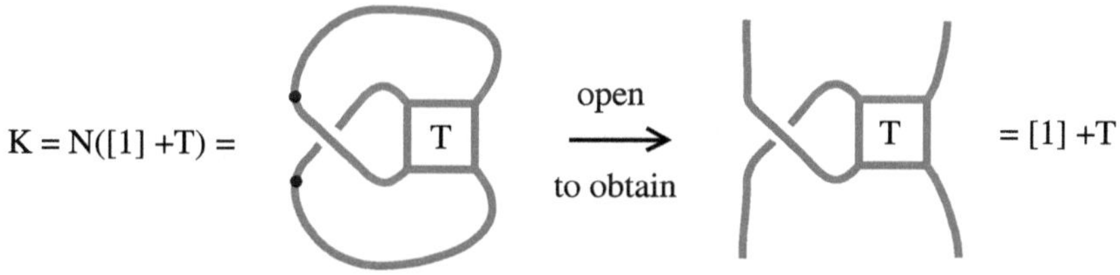

Fig. 5.21. A standard cut

has fraction P/Q, for $|P| > |Q|$. Thus R can be written in the form $R = [1]+T$ or $R = [-1]+T$. We consider the rational knot diagram $K = N([1]+T)$, see Fig. 5.21. (We analyse $N([-1]+T)$ in the same way.) The tangle $[1]+T$ is said to arise as a *standard cut* on K.

Notice that the indicated horizontal crossing of $N([1]+T)$ could also be seen as a vertical one. So we could also cut the diagram K at the two other marked points (see Fig. 5.22) and still obtain a rational tangle, since T is rational. The tangle obtained by cutting K in this second pair of points is said to arise as a *special cut* on K. Figure 5.22 demonstrates that the tangle of the special cut is the tangle $[-1]-1/T$. So we have $N([1]+T) \sim N([-1]-1/T)$. Suppose now $F(T) = p/q$. Then $F([1]+T) = 1+p/q = (p+q)/q$, while $F([-1]-1/T) = -1-q/p = (p+q)/(-p)$, so the two rational tangles that give rise to the same knot K are not isotopic. Since $-p \equiv q \bmod (p+q)$, this equivalence is another example for Theorem 2. In Fig. 5.22 if we took $T = 1/[2]$ then $[-1]-1/T = [-3]$ and we would obtain the example of Fig. 5.20.

The proof of Theorem 2 can now proceed in two stages. First, given a rational knot diagram we look for all possible places where we could cut and open it to a rational tangle. The crux of our proof in [17] is the fact that all possible "rational cuts" on a rational knot fall into one of the basic cases that we have already discussed, i.e. we have the standard cuts, the palindrome cuts and the special cuts. In Fig. 5.23 we illustrate on a representative rational knot, all the cuts that exhibit that knot as a closure of a rational tangle. Each pair of points is marked with the same number. The arithmetic is similar to

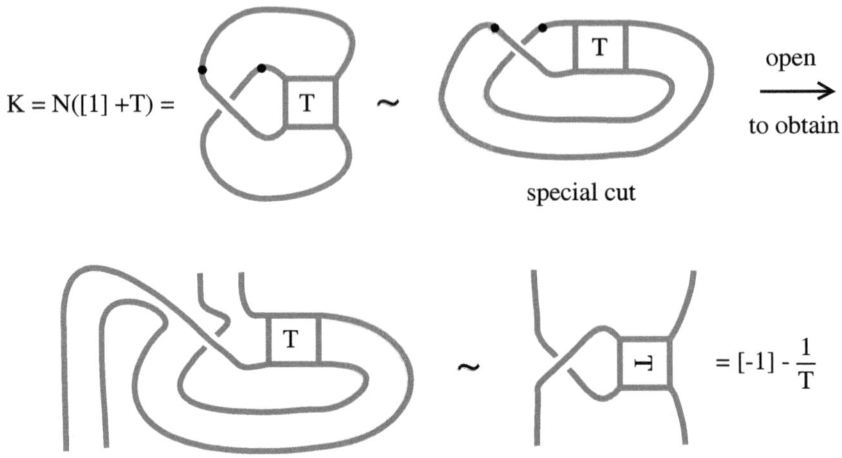

Fig. 5.22. A special cut

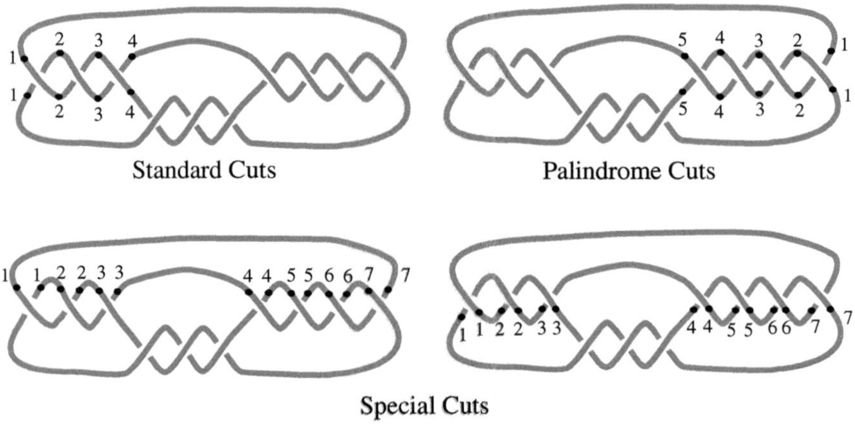

Fig. 5.23. Standard, palindrome and special cuts

the cases that have been already verified. It is convenient to say that reduced fractions p/q and p'/q' are *arithmetically equivalent*, written $p/q \sim p'/q'$ if $p = p'$ and either $qq' \equiv 1 \pmod{p}$ or $q \equiv q' \pmod{p}$. In this language, Schubert's theorem states that two rational tangles close to form isotopic knots if and only if their fractions are arithmetically equivalent.

In Fig. 5.24 we illustrate one example of a cut that is not allowed since it opens the knot to a non-rational tangle.

In the second stage of the proof we want to check the arithmetic equivalence for two different given knot diagrams, numerators of some rational tangles. By Lemma 2 the two knot diagrams may be assumed alternating, so

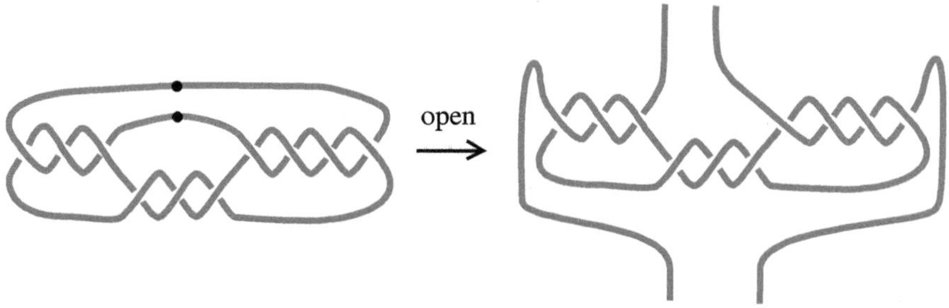

Fig. 5.24. A non-rational cut

by the Tait conjecture they will differ by flypes. We analyse all possible flypes to prove that no new cases for study arise. Hence the proof becomes complete at that point. We refer the reader to our study [17] for details.

Remark 3 The original proof of the classification of unoriented rational knots by Schubert [14] proceeded by a different route than the proof we have just sketched. Schubert used a two-bridge representation of rational knots (representing the knots and links as diagrams in the plane with two special overcrossing arcs, called the bridges). From the two-bridge representation, one could extract a fraction p/q, and Schubert showed by means of a canonical form, that if two such presentations are isotopic, then their fractions are arithmetically equivalent (in the sense that we have described here). On the other hand, Seifert [14] observed that the twofold branched covering space of a two-bridge presentation with fraction p/q is a lens space of type $L(p,q)$. Lens spaces are a particularly tractable set of three manifolds, and it is known by work of Reidemeister and Franz [15, 34] that $L(p,q)$ is homeomorphic to $L(p',q')$ if and only if p/q and p'/q' are arithmetically equivalent. Furthermore, one knows that if knots K and K' are isotopic, then their twofold branched covering spaces are homeomorphic. Hence it follows that if two rational knots are isotopic, then their fractions are arithmetically equivalent (via the result of Reidemeister and Franz classifying lens spaces). In this way Schubert proved that two rational knots are isotopic if and only if their fractions are arithmetically equivalent.

5.6 Rational Knots and Their Mirror Images

In this section we give an application of Theorem 2. An unoriented knot or link K is said to be *achiral* if it is topologically equivalent to its mirror image $-K$. If a link is not equivalent to its mirror image then it is said be *chiral.* One then can speak of the *chirality* of a given knot or link, meaning whether it is chiral or achiral. Chirality plays an important role in the applications of knot theory to chemistry and molecular biology. It is interesting to use the classification of rational knots and links to determine their chirality. Indeed, we have the following well-known result (for example see [5] and also page 24, Exercise 2.1.4 in [9]):

Theorem 4. *Let $K = N(T)$ be an unoriented rational knot or link, presented as the numerator of a rational tangle T. Suppose that $F(T) = p/q$ with p and q relatively prime. Then K is achiral if and only if $q^2 \equiv -1 \pmod{p}$. It follows that achiral rational knots and links are all numerators of rational tangles of the form $[[a_1],[a_2],\ldots,[a_k],[a_k],\ldots,[a_2],[a_1]]$ for any integers $a_1,\ldots,a_k$.*

Note that in this description we are using a representation of the tangle with an even number of terms. The leftmost twists $[a_1]$ are horizontal, thus the rightmost starting twists $[a_1]$ are vertical.

Proof. With $-T$ the mirror image of the tangle T, we have that $-K = N(-T)$ and $F(-T) = p/(-q)$. If K is topologically equivalent to $-K$, then $N(T)$ and $N(-T)$ are equivalent, and it follows from the classification theorem for rational knots that either $q(-q) \equiv 1 \pmod{p}$ or $q \equiv -q \pmod{p}$. Without loss of generality we can assume that $0 < q < p$. Hence $2q$ is not divisible by p and therefore it is not the case that $q \equiv -q \pmod{p}$. Hence $q^2 \equiv -1 \pmod{p}$.

Conversely, if $q^2 \equiv -1 \pmod{p}$, then it follows from the palindrome theorem (described in the previous section) [17] that *the continued fraction expansion of p/q has to be symmetric with an even number of terms.* It is then easy to see that the corresponding rational knot or link, say $K = N(T)$, is equivalent to its mirror image. One rotates K by 180° in the plane and swings an arc, as Fig. 5.25 illustrates. This completes the proof.

In [35] the authors find an explicit formula for the number of achiral rational knots among all rational knots with n crossings.

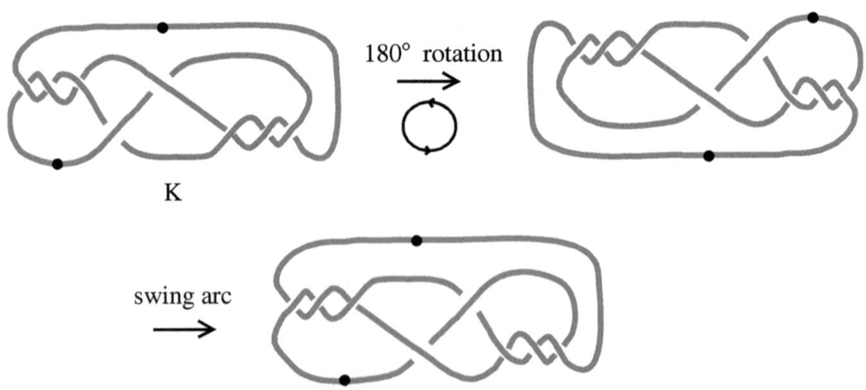

Fig. 5.25. An achiral rational link

5.7 The Oriented Case

Oriented rational knots and links arise as numerator closures of oriented rational tangles. In order to compare oriented rational knots via rational tangles we need to examine how rational tangles can be oriented. We orient rational tangles by choosing an orientation for each strand of the tangle. Here we are only interested in orientations that yield consistently oriented knots upon taking the numerator closure. This means that the two top end arcs have to be oriented one inward and the other outward. Same for the two bottom end arcs. We shall say that two oriented rational tangles are *isotopic* if they are isotopic as unoriented tangles, by an isotopy that carries the orientation of one tangle to the orientation of the other. Note that, since the end arcs of a tangle are fixed during a tangle isotopy, this means that the tangles must have identical orientations at their four end arcs *NW, NE, SW, SE.* It follows that if we change the orientation of one or both strands of an oriented rational tangle we will always obtain a non-isotopic oriented rational tangle.

Reversing the orientation of one strand of an oriented rational tangle may or may not give rise to isotopic oriented rational knots. Figure 5.26 illustrates an example of non-isotopic oriented rational knots, which are isotopic as unoriented knots.

Reversing the orientation of both strands of an oriented rational tangle will always give rise to two isotopic oriented rational knots or links. We can see this by doing a vertical flip, as Fig. 5.27 demonstrates. Using this observation we conclude that, as far as the study of oriented rational knots is concerned, *all oriented rational tangles may be assumed to have the same orientation for*

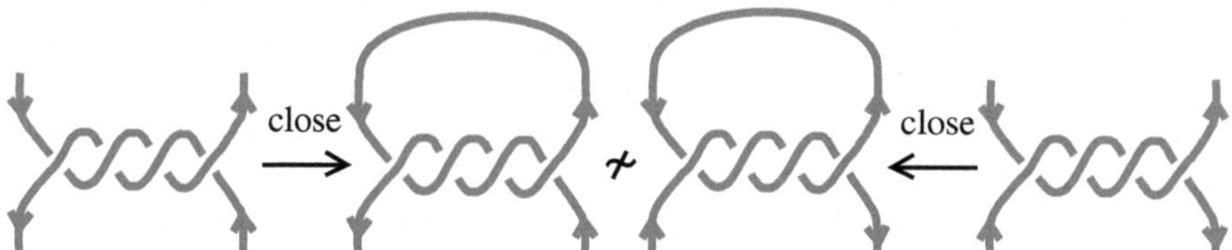

Fig. 5.26. Non-isotopic oriented rational Links

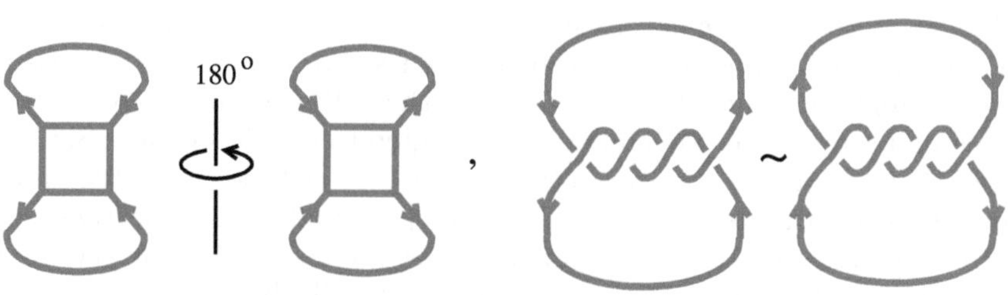

Fig. 5.27. Isotopic oriented rational knots and links

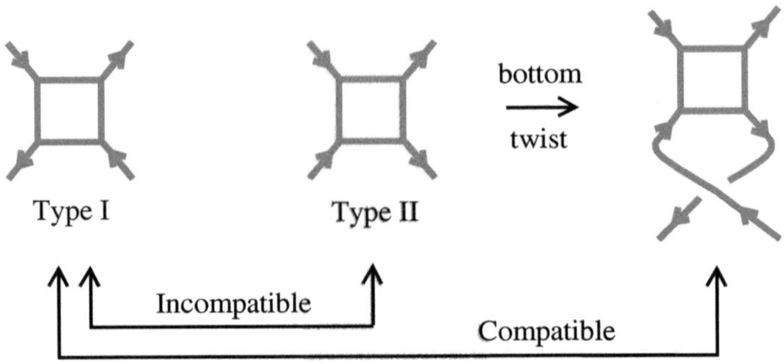

Fig. 5.28. Compatible and incompatible orientations

their NW and NE end arcs. We fix this orientation to be downward for the *NW* end arc and upward for the *NE* arc, as in the examples of Fig. 5.26 and as illustrated in Fig. 5.28. Indeed, if the orientations are opposite of the fixed ones doing a vertical flip the knot may be considered as the numerator of the vertical flip of the original tangle. But this is unoriented isotopic to the original tangle (recall Sect. 5.2, Fig. 5.7), whilst its orientation pattern agrees with our convention.

Thus we reduce our analysis to two basic types of orientation for the four end arcs of a rational tangle. We shall call an oriented rational tangle of *type I* if the *SW* arc is oriented upwards and the *SE* arc is oriented downwards, and of *type II* if the *SW* arc is oriented downward and the *SE* arc is oriented upward, see Fig. 5.28. From the above remarks, any tangle is of type I or type II. Two tangles are said to be *compatible* it they are both of type I or both of type II and *incompatible* if they are of different types. In order to classify oriented rational knots seen as numerator closures of oriented rational tangles, we will always compare compatible rational angles. Note that if two oriented tangles are incompatible, adding a single half twist at the bottom of one of them yields a new pair of compatible tangles, as Fig. 5.28 illustrates. Note also that adding such a twist, although it changes the tangle, it does not change the isotopy type of the numerator closure. Thus, up to bottom twists, we are always able to compare oriented rational tangles of the same orientation type.

We now introduce the notion of *connectivity* and we relate it to orientation and the fraction of unoriented rational tangles. We say that an unoriented rational tangle has *connectivity* type $[0]$ if the *NW* end arc is connected to the *NE* end arc and the *SW* end arc is connected to the *SE* end arc. Similarly, we say that the tangle has *connectivity* type $[+1]$ or type $[\infty]$ if the end arc connections are the same as in the tangles $[+1]$ and $[\infty]$, respectively. The basic connectivity patterns of rational tangles are exemplified by the tangles $[0]$, $[\infty]$ and $[+1]$. We can represent them iconically by the symbols shown below.

$$[0] = \asymp$$
$$[\infty] = ><$$
$$[+1] = \chi.$$

Note that connectivity type $[0]$ yields two-component rational links, while type $[+1]$ or $[\infty]$ yields one-component rational links. Also, adding a bottom twist to a rational tangle of connectivity type $[0]$ will not change the connectivity type of the tangle, while adding a bottom twist to a rational tangle of connectivity type $[\infty]$ will switch the connectivity type to $[+1]$ and vice versa. While the connectivity type of unoriented rational tangles may be $[0]$, $[+1]$ or $[\infty]$, note that an oriented rational tangle of type I will have connectivity type $[0]$ or $[\infty]$ and an oriented rational tangle of type II will have connectivity type $[0]$ or $[+1]$.

Further, we need to keep an accounting of the connectivity of rational tangles in relation to the parity of the numerators and denominators of their fractions. We refer the reader to our Study [17] for a full account.

We adopt the following notation: e stands for *even* and o stands for *odd*. The *parity* of a fraction p/q is defined to be the ratio of the parities (e or o) of its numerator and denominator p and q. Thus the fraction $2/3$ is of parity e/o. The tangle $[0]$ has fraction $0 = 0/1$, thus parity e/o; the tangle $[\infty]$ has fraction $\infty = 1/0$, thus parity o/e; and the tangle $[+1]$ has fraction $1 = 1/1$, thus parity o/o. We then have the following result.

Theorem 5. *A rational tangle T has connectivity type $\asymp$ if and only if its fraction has parity e/o. T has connectivity type $><$ if and only if its fraction has parity o/e. T has connectivity type χ if and only if its fraction has parity o/o. (Note that the formal fraction of $[\infty]$ itself is $1/0$.) Thus the link $N(T)$ has two components if and only if T has fraction $F(T)$ of parity e/o.*

We will now proceed with sketching the proof of Theorem 3. We shall prove Schubert's oriented theorem by referring to our previous work on the unoriented case and then analyzing how orientations and fractions are related. Our strategy is as follows: Consider an oriented rational knot or link diagram K in the form $N(T)$, where T is a rational tangle in continued fraction form. Then any other rational tangle that closes to this knot $N(T)$ is available, up to bottom twists if necessary, as a cut from the given diagram. If two rational tangles close to give K as an unoriented rational knot or link, then there are orientations on these tangles, induced from K so that the oriented tangles close to give K as an oriented knot or link. The two tangles may or may not be compatible. Thus, we must analyze when, comparing with the standard cut for the rational knot or link, another cut produces a compatible or incompatible rational tangle. However, assuming the top orientations are the same, we can replace one of the two incompatible tangles by the tangle obtained by adding a twist at the bottom. *It is this possible twist difference that gives rise to the change from modulus p in the unoriented case to the modulus 2p in the oriented case.* We now perform this analysis. There are many interesting aspects to this analysis and we refer the reader to our study [17] for these details. Schubert [14] proved his version of the oriented theorem by using the two-bridge representation of rational knots and links, see also [6]. We give a tangle-theoretic combinatorial proof based upon the combinatorics of the unoriented case.

The simplest instance of the classification of oriented rational knots is adding an *even number of twists* at the bottom of an oriented rational tangle T, see Fig. 5.28. We then obtain a compatible tangle $T * 1/[2n]$, and $N(T * 1/[2n]) \sim N(T)$. If now $F(T) = p/q$, then $F(T*1/[2n]) = F(1/([2n]+1/T)) = 1/(2n + 1/F(T)) = p/(2np + q)$. Hence, if we set $2np + q = q'$ we have $q \equiv q' (\mathrm{mod}\, 2p)$, just as the oriented Schubert theorem predicts. Note that reducing all possible bottom twists implies $|p| > |q|$ for both tangles, if the two tangles that we compare each time are compatible or for only one, if they are incompatible.

We then have to compare the special cut and the palindrome cut with the standard cut. In the oriented case the special cut is easier to see whilst the palindrome cut requires a more sophisticated analysis. Figure 5.29 illustrates the general case of the special cut. In order to understand Fig. 5.29 it is necessary to also view Fig. 5.22 for the details of this cut.

Recall that if $S = [1] + T$ then the tangle of the special cut on the knot $N([1] + T)$ is the tangle $S' = [-1] - 1/T$. And if $F(T) = p/q$ then $F([1] + T) = (p + q)/q$ and $F([-1] - 1/T) = (p + q)/ - p$. Now, the point is that the orientations of the tangles S and S' are incompatible. Applying a $[+1]$ bottom twist to S' yields $S'' = ([-1]1/T) * [1]$, and we find that $F(S'') = (p+q)/q$. Thus, the oriented rational tangles S and S'' have the same fraction and by Theorem 1 and their compatibility they are oriented isotopic and the arithmetics of Theorem 3 is straightforward.

We are left to examine the case of the palindrome cut. For this part of the proof, we refer the reader to our study [17].

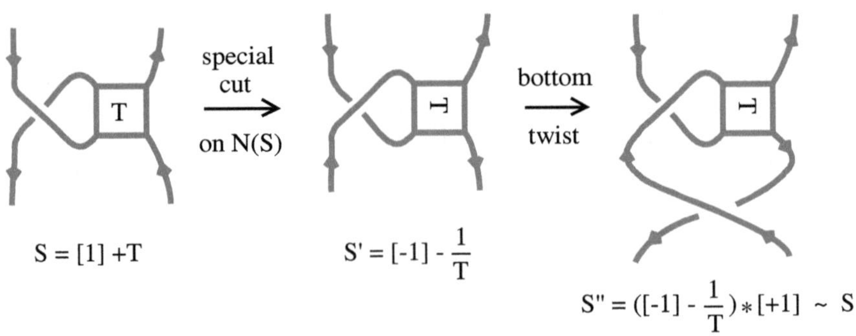

Fig. 5.29. The oriented special cut

5.8 Strongly Invertible Links

An oriented knot or link is invertible if it is oriented isotopic to the link obtained from it by reversing the orientation of each component. We have seen (see Fig. 5.27) that rational knots and links are invertible. A link L of two components is said to be *strongly invertible* if L is ambient isotopic to itself with the orientation of only one component reversed. In Fig. 5.30 we illustrate the link $L = N([[2],[1],[2]])$. This is a strongly invertible link as is apparent by a 180° vertical rotation. This link is well known as the Whitehead link, a link with linking number zero. Note that since $[[2],[1],[2]]$ has fraction equal to $1 + 1/(1 + 1/2) = 8/3$ this link is non-trivial via the classification of rational knots and links. Note also that $3 \cdot 3 = 1 + 1 \cdot 8$.

In general we have the following. For our proof, see [17].

Theorem 6. *Let $L = N(T)$ be an oriented rational link with associated tangle fraction $F(T) = p/q$ of parity e/o, with p and q relatively prime and $|p| > |q|$. Then L is strongly invertible if and only if $q^2 = 1 + up$ with u an odd integer. It follows that strongly invertible links are all numerators of rational tangles of the form $[[a_1],[a_2],\ldots,[a_k],[\alpha],[a_k],\ldots,[a_2],[a_1]]$ for any integers $a_1,\ldots,a_k,\alpha$.*

(See Fig. 5.31 for another example of a strongly invertible link.) In this case the link is $L = N([[3],[1],[1],[1],[3]])$ with $F(L) = 40/11$. Note that $11^2 = 1 + 3 \cdot 40$, fitting the conclusion of Theorem 6.

5.9 Applications to the Topology of DNA

DNA supercoils, replicates and recombines with the help of certain enzymes. *Site-specific recombination* is one of the ways nature alters the genetic code of an organism, either by moving a block of DNA to another position on the

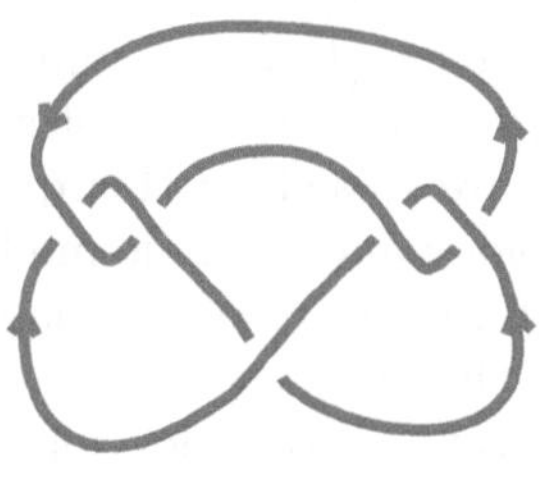

N([[2], [1], [2]]) = W
the Whitehead Link
F(W) = 2+1/(1+1/2) = 8/3
3·3 = 1 + 1·8

Fig. 5.30. The whitehead link is strongly invertible

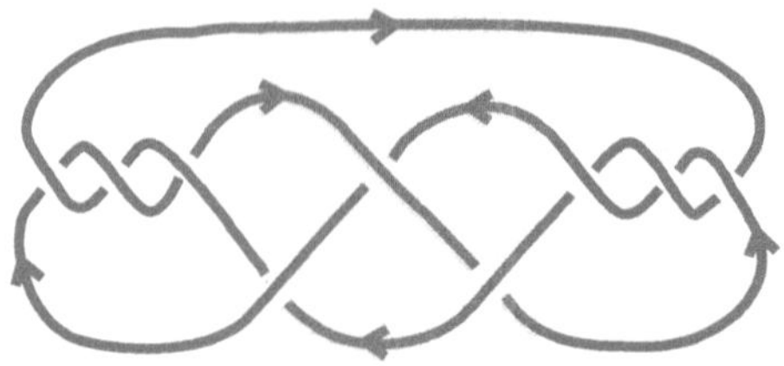

Fig. 5.31. An example of a strongly invertible link

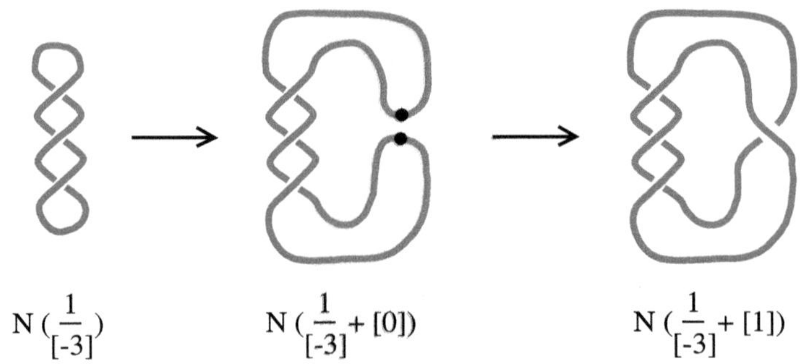

Fig. 5.32. Global picture of recombination

molecule or by integrating a block of alien DNA into a host genome. For a closed molecule of DNA a global picture of the recombination would be as shown in Fig. 5.32, where double-stranded DNA is represented by a single line and the recombination sites are marked with points. This picture can be interpreted as $N(S+[0]) \longrightarrow N(S+[1])$, for $S = 1/[-3]$ in this example. This operation can be repeated as in Fig. 5.33. Note that the $[0] - [\infty]$ interchange of Fig. 5.10 can be seen as the first step of the process.

In this depiction of recombination, we have shown a local replacement of the tangle $[0]$ by the tangle $[1]$ connoting a new cross-connection of the DNA strands. In general, it is not known without corroborating evidence just what the topological geometry of the recombination replacement will be. Even in the case of a single half-twist replacement such as $[1]$, it is certainly not obvious beforehand that the replacement will always be $[+1]$ and not sometimes the reverse twist of $[-1]$. It was at the juncture raised by this question that a combination of topological methods in biology and a tangle model using knot theory developed by C. Ernst and D.W. Sumners resolved the issue in some specific cases. (See [36, 37] and references therein.)

On the biological side, methods of protein coating developed by N. Cozzarelli, S.J. Spengler and A. Stasiak et al. in [38] made it possible for the first time to see knotted DNA in an electron micrograph with sufficient resolution

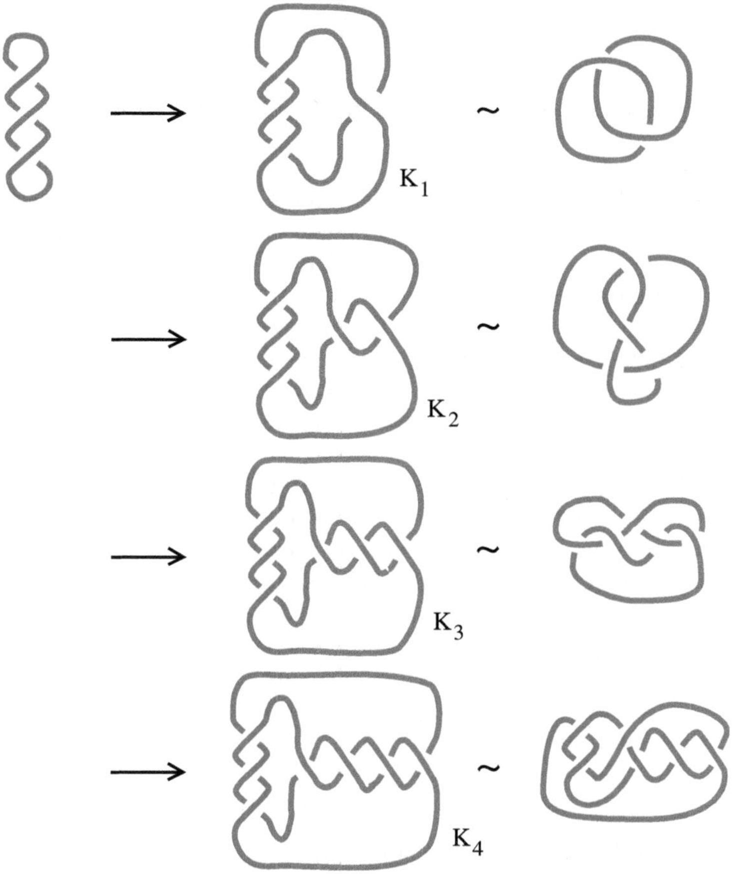

Fig. 5.33. Multiple recombinations

to actually identify the topological type of these knots. The protein coating technique made it possible to design an experiment involving successive DNA recombinations and to examine the topology of the products. In [38] the knotted DNA produced by such successive recombinations was consistent with the hypothesis that all recombinations were of the type of a positive half twist as in [+1]. Then D.W. Sumners and C. Ernst [36] proposed a *tangle model for successive DNA recombinations* and showed, in the case of the experiments in question, that there was no other topological possibility for the recombination mechanism than the positive half-twist [+1]. This constituted a unique use of topology as a theoretical underpinning for a problem in molecular biology.

Here is a brief description of the tangle model for DNA recombination. It is assumed that the initial state of the DNA is described as the numerator closure $N(S)$ of a *substrate tangle* S. The local geometry of the recombination is assumed to be described by the replacement of the tangle [0] with a specific tangle R. The results of the successive rounds of recombination are the knots and links

$$N(S+R)=K_1, \quad N(S+R+R)=K_2, \quad N(S+R+R+R)=K_3, \quad \ldots$$

Knowing the knots $K_1, K_2, K_3, \ldots$ one would like to solve the above system of equations with the tangles S and R as unknowns.

For such experiments Ernst and Sumners [36] used the classification of rational knots in the unoriented case, as well as results of Culler, Gordon, Luecke and Shalen [39] on Dehn surgery to prove that the solutions $S+nR$ must be *rational tangles.* These results of Culler, Gordon, Luecke and Shalen show the topologist under what circumstances a three-manifold with cyclic fundamental group must be a lens space. By showing when the twofold branched covers of the DNA knots must be lens spaces, the recombination problems are reduced to the consideration of rational knots. This is a deep application of the three-manifold approach to rational knots and their generalizations.

One can then apply the theorem on the classification of rational knots to deduce (in these instances) the uniqueness of S and R. Note that, in these experiments, the substrate tangle S was also pinpointed by the sequence of knots and links that resulted from the recombination.

Here we solve tangle equations like the above under rationality assumptions on all tangles in question. This allows us to use only the mathematical techniques developed in this chapter. We shall illustrate how a sequence of rational knots and links

$$N(S+nR)=K_n, \quad n=0,1,2,3,\ldots$$

with S and R rational tangles, such that $R=[r]$, $F(S)=p/q$ and $p, q, r \in \mathbb{Z}$ $(p>0)$ *determines* p/q *and* r *uniquely* if we know sufficiently many K_n. We call this the "DNA knitting machine analysis".

Theorem 7. *Let a sequence* K_n *of rational knots and links be defined by the equations* $K_n = N(S+nR)$ *with specific integers* p, q, r $(p>0)$, *where* $R=[r]$, $F(S)=p/q$. *Then* p/q *and* r *are uniquely determined if one knows the topological type of the unoriented links* $K_0, K_1, \ldots, K_N$ *for any integer* $N \geq |q| - p/qr$.

Proof. In this proof we write $N(p/q+nr)$ or $N(p+qnr/q)$ for $N(S+nR)$. We shall also write $K = K'$ to mean that K and K' are isotopic links. Moreover we shall say for a pair of reduced fractions P/q and P/q' that q and q' are *arithmetically related relative to* P if either $q \equiv q' (\operatorname{mod} P)$ or $qq' \equiv 1 (\operatorname{mod} P)$. Suppose the integers p, q, r give rise to the sequence of links $K_0, K_1, \ldots$ Suppose there is some other triple of integers p', q', r' that give rise to the same sequence of links. We will show uniqueness of p, q, r under the conditions of the theorem. We shall say "the equality holds for n" to mean that $N((p+qrn)/q) = N((p'+q'r'n)/q')$. We suppose that $K_n = N((p+qrn)/q)$ as in the hypothesis of the theorem, and suppose that there are p', q', r' such that for some n (or a range of values of n to be specified below) $K_n = N((p' + q'r'n)/q')$.

If $n = 0$ then we have $N(p/q) = N(p'/q')$. Hence by the classification theorem we know that $p = p'$ and that q and q' are arithmetically related. Note that the same argument shows that if the equality holds for any two consecutive values of n, then $p = p'$. We shall assume henceforth that $p = p'$. With this assumption in place, we see that if the equality holds for any $n \neq 0$ then $qr = q'r'$. Hence we shall assume this as well from now on.

If $|p + qrn|$ is sufficiently large, then the congruences for the arithmetical relation of q and q' must be *equalities over the integers.* Since $qq' = 1$ over the integers can hold only if $q = q' = 1$ or -1 we see that it must be the case that $q = q'$ if the equality is to hold for sufficiently large n. From this and the equation $qr = q'r'$ it follows that $r = r'$. It remains to determine a bound on n. In order to be sure that $|p + qrn|$ is sufficiently large, we need that $|qq'| \leq |p + qrn|$. Since $q'r' = qr$, we also know that $|q'| \leq |qr|$. Hence n is sufficiently large if $|q^2r| \leq |p + qrn|$.

If $qr > 0$ then, since $p > 0$, we are asking that $|q^2r| \leq p + qrn$. Hence

$$n \geq (|q^2r| - p)/(qr) = |q| - (p/qr).$$

If $qr < 0$ then for n large we will have $|p+qrn| = -p-qrn$. Thus we want to solve $|q^2r| \leq -p - qrn$, whence

$$n \geq (|q^2r| + p)/(-qr) = |q| - (p/qr).$$

Since these two cases exhaust the range of possibilities, this completes the proof of the theorem.

Here is a special case of Theorem 7. (See Fig. 5.33.) Suppose that we were given a sequence of knots and links K_n such that

$$K_n = N\left(\frac{1}{[-3]} + [1] + [1] + \ldots + [1]\right) = N\left(\frac{1}{[-3]} + n\,[1]\right).$$

We have $F(1/[-3] + n\,[1]) = (3n - 1)/3$ and we shall write $K_n = N([(3n - 1)/3])$. We are told that each of these rational knots is in fact the numerator closure of a rational tangle denoted

$$[p/q] + n\,[r]$$

for some rational number p/q and some integer r. That is, we are told that they come from a DNA knitting machine that is using rational tangle patterns. But we only know the knots and the fact that they are indeed the closures for $p/q = -1/3$ and $r = 1$. By this analysis, the uniqueness is implied by the knots and links $\{K_1, K_2, K_3, K_4\}$. This means that a DNA knitting machine $K_n = N(S + nR)$ that emits the four specific knots $K_n = N([(3n-1)/3])$ for $n = 1, 2, 3, 4$ must be of the form $S = 1/[-3]$ and $R = [1]$. It was in this way (with a finite number of observations) that the structure of recombination in T_{n3} resolvase was determined [37].

In this version of the tangle model for DNA recombination we have made a blanket assumption that the substrate tangle S and the recombination tangle R and all the tangles $S + nR$ were rational. Actually, if we assume that S is rational and that $S + R$ is rational, then it follows that R is an integer tangle. Thus S and R neccessarily form a DNA knitting machine under these conditions. It is relatively natural to assume that S is rational on the grounds of simplicity. On the other hand it is not so obvious that the recombination tangle should be an integer. The fact that the products of the DNA recombination experiments yield rational knots and links, lends credence to the hypothesis of rational tangles and hence integral recombination tangles. But there certainly is a subtlety here, since we know that the numerator closure of the sum of two rational tangles is always a rational knot or link. In fact, it is here that some deeper topology shows that certain rational products from a generalized knitting machine of the form $K_n = N(S + nR)$, where S and R are arbitrary tangles will force the rationality of the tangles $S + nR$. We refer the reader to [36, 40, 41] for details of this approach.

References

1. M.M. Asaeda, J.H. Przytycki, A.S. Sikora, Kauffman-Harary conjecture holds for Montesinos knots. J. Knot Theory Ramifications **13** (4), 467–477 (2004)
2. C. Bankwitz, H.G. Schumann, Abh. Math. Sem. Univ. Hamburg **10**, 263–284 (1934)
3. S.A. Bleiler, Y.H. Moriah, Splittings and branched coverings of B^3, Math. Ann. **281** (4), 531–543 (1988)
4. J.H. Conway, An enumeration of knots and links and some of their algebraic properties, *Proceedings of the Conference on Computational Problems in Abstract Algebra Held at Oxford in 1967*, J. Leech ed., (First edition 1970), (Pergamon Press, 1970), pp. 329–358
5. L. Siebenmann, Lecture Notes on Rational Tangles, Orsay (1972) (unpublished)
6. G. Burde, H. Zieschang, "Knots", de Gruyter Studies in Mathematics **5** (1985)
7. L.H. Kauffman, Knot Logic, *Knots and Applications*, Series on Knots and Everything, **2**, L.H. Kauffman ed., World Scientific, (1995)
8. J.R. Goldman, L.H. Kauffman, Adv. Appl. Math. **18**, 300–332 (1997)
9. A. Kawauchi, '*A Survey of Knot Theory*' (Birkhäuser, Verlag, 1996)
10. W.B.R. Lickorish, '*An Introduction to Knot Theory*', Springer Graduate Texts in Mathematics **175** (1997)
11. K. Murasugi, '*Knot Theory and its Applications*', Translated from the 1993 Japanese original by B. Kurpita, Birkhäuser Verlag (1996)
12. L.H. Kauffman, Topology, **26**, 395–407 (1987)
13. L.H. Kauffman, S. Lambropoulou, On the classification of rational tangles, to appear in Advances in Applied Math. (See http://www.math.uic.edu/~kauffman/ or http://users.ntua.gr/sofial or math.GT/0311499)
14. H. Schubert, *Math. Zeitschrift* **65**, 133–170 (1956)
15. K. Reidemeister, Abh. Math. Sem. Hansischen Univ. **11**, 102–109 (1936)
16. G. Burde, Math. Zeitschrift **145**, 235–242 (1975)

17. L.H. Kauffman, S. Lambropoulou, On the classification of rational knots, to appear in *L' Enseignement Math.* (See http://www.math.uic.edu/~kauffman/ or http://users.ntua.gr/sofial or math.GT/0212011)
18. K. Reidemeister, "Knotentheorie" (Reprint), Chelsea, New York (1948)
19. P.G. Tait, On knots, I, II, III, Scientific Papers, **1**, Cambridge University Press, Cambridge, 273–347 (1898)
20. W. Menasco, M. Thistlethwaite, Annals of Mathematics **138**, 113–171 (1993)
21. A.Ya. Khinchin, '*Continued Fractions*', Dover (republication of the 1964 edition of Chicago Univ. Press) (1997)
22. C.D. Olds, '*Continued Fractions*', New Mathematical Library, Math. Assoc. of Amerika, **9** (1963)
23. K. Kolden, Arch. Math. og Naturvidenskab **6**, 141–196 (1949)
24. L.H. Kauffman, Ann. Math. Stud. **115** (Princeton Univ. Press, Princeton, NJ, 1987)
25. L.H. Kauffman, S. Lambropoulou, On the classififcation of rational tangles. Adv. in Appln. Math. **33** (2), 199–237 (2004)
26. L.H. Kauffman, S. Lambropoulou, On the classififcation of rational knots. Enseign. Math. (2) **49** (3–4) 357–410 (2003)
27. V.F.R. Jones, Bull. Am. Math. Soc. (N.S.) **12** (1), 103–111 (1985)
28. V.F.R. Jones, Notices Am. Math. Soc. **33** (2), 219–225 (1986)
29. D.A. Krebes, J. Knot Theory Ramifications **8** (3), 321–352 (1999)
30. L.H. Kauffman, "*Formal Knot Theory*", Mathematical Notes **30**, Princeton Univ. Press, Princeton, NJ, (1983)
31. L. Person, M. Dunne, J. DeNinno, B. Guntel and L. Smith, Colourings of rational, alternating knots and links, (preprint 2002)
32. L.H. Kauffman, F. Harary, Knots and Graphs I – Arc Graphs and Colourings, Advances in Applied Mathematics **22**, 312–337 (1999)
33. M. Asaeda, J. Przytycki, A. Sikora, Kauffman–Harary Conjecture holds for Montesinos Knots (to appear in JKTR)
34. W. Franz, J. Reine Angew. Math. **173**, 245–254 (1935)
35. C. Ernst, D.W. Sumners, Math. Proc. Camb. Phil. Soc. **102**, 303–315 (1987)
36. C. Ernst, D.W. Sumners, Math. Proc. Camb. Phil. Soc. **108**, 489–515 (1990)
37. D.W. Sumners, Math. Intelligencer **12**, 71–80 (1990)
38. N. Cozzarelli, F. Dean, T. Koller, M.A. Krasnow, S.J. Spengler and A. tasiak Nature **304**, 550–560 (1983)
39. M.C. Culler, C.M. Gordon, J. Luecke and P.B. Shalen, Ann. Math. **125**, 237–300 (1987)
40. C. Ernst, D.W. Sumners, Math. Proc. Cambridge Philos. Soc. **126** (1), 23–36 (1999)
41. I.K. Darcy, Solving unoriented tangle equations involving 4-plats. J. Knot Theory Ramifications **14** (8), 993–1005 (2005)
42. E.J. Brody, Ann. Math. **71**, 163–184 (1960)
43. J.M. Montesinos, Revetements ramifies des noeuds, Espaces fibres de Seifert et scindements de Heegaard, *Publicaciones del Seminario Mathematico Garcia de Galdeano, Serie II, Seccion 3* (1984)
44. V.V. Prasolov, A.B. Sossinsky, "Knots, Links, Braids and 3-Manifolds", AMS Translations of Mathematical Monographs **154** (1997)
45. K. Reidemeister Abh. Math. Sem. Univ. Hamburg **5**, 24–32 (1927)
46. D. Rolfsen, '*Knots and Links*' (Publish or Perish Press, Berkeley 1976)

47. H. Seifert, Abh. Math. Sem. Univ. Hamburg, **11**, 84–101 (1936)
48. J. Sawollek, Tait's flyping conjecture for 4-regular graphs, preprint (1998)
49. C. Sundberg, M. Thistlethwaite, The rate of growth of the number of alternating links and tangles, Pacific J. Math. **182**(2), 329–358 (1998)
50. H.S. Wall, '*Analytic Theory of Continued Fractions*' (D. Van Nostrand Company, Inc., 1948)

6

Linear Behavior of the Writhe Versus the Number of Crossings in Rational Knots and Links

C. Cerf and A. Stasiak

Summary. Using the formula introduced in [Proc. Natl Acad. Sci. USA **97**, 3795 (2000)], we can predict the 3D writhe of any rational knot or link in its ideal configuration, or equivalently, the ensemble average of the 3D writhe of random configurations of it. Here we present a method that allows us to express the writhe as a linear function of the minimal crossing number within individual Conway families of rational knots and links. We discuss the cases of families with slopes $\pm 4/7, \pm 10/7, \pm 1$, and 0. For families with the same slope value, the vertical shift between the corresponding lines can also be computed.

6.1 Introduction

Quantization of writhe in knots is a puzzling phenomenon, which was initially discussed only among a narrow group of specialists but recently became quite famous [1]. Let us explain what this concept covers. Writhe (or 3D writhe, or Wr) is a measure of chirality of oriented closed curves in 3D space. It corresponds to the average signed number of perceived self-crossings in an oriented curve when observed from a random direction, where each right-handed crossing is scored as $+1$ and each left-handed crossing is scored as -1. Writhe is usually calculated using a Gauss integral formula [2].

Studies of random walks in a cubic lattice revealed that, while different realizations of random knots of a given type have stochastically distributed values of their writhe, the average of writhe over the statistical ensemble of knots of a given type, like right-handed trefoils for example, reaches a characteristic value that is independent of the length of the walk [3, 4]. Thus for example the average writhe of random right-handed trefoils in a cubic lattice does not change when the number of segments is increased from 34 to 250. The same studies pointed out that there is a constant increase (i.e., a *quantization*) of the average writhe between successive knots belonging to families of torus knots like $3_1, 5_1, 7_1, 9_1$, etc. or to twist knots like $4_1, 6_1, 8_1$, etc. The specific difference of average writhe between successive knots belonging to different families depends on the particular family of knots.

Studies of random knots that are not confined to a lattice also revealed the same phenomenon. The average writhe over a statistical ensemble of random walks not confined to a lattice but forming a given knot type was practically the same as the average writhe of a statistical ensemble of random knots of the same type in a cubic lattice [2, 4]. The average writhe over a statistical ensemble of simulated configurations of a given knot closely corresponds to the time-averaged writhe of a randomly fluctuating long polymeric chain forming the same knot type. Therefore a freely fluctuating long DNA molecule closed into a right-handed trefoil, for example, would show the same average writhe as a freely fluctuating long polyethylene molecule that is also closed into a right-handed trefoil. The time-averaged values of writhe seem to be independent of the size of the polymeric chain [2]. In addition the differences of time-averaged writhe between freely fluctuating successive knots belonging to a given family seem to be constant [2].

The quantization of writhe observed for statistical ensembles of different knots is mysteriously reflected by the quantization of writhe for the so-called *ideal knots*. Ideal knots are defined as shortest possible paths of cylindrical tubes with uniform diameter that can still be closed into a given knot type [2]. Numerical simulations revealed that writhe of axial trajectories of unique realizations of ideal knots of a given type corresponds to the time-averaged writhe of freely fluctuating knots of the same type [2]. Therefore unique representations of ideal knots of a given type "capture" the essential statistical property of random knots of a given type [5]. To find the average writhe of fluctuating knots of a given type it becomes therefore much more practical to measure the writhe of one ideal configuration of this knot instead of simulating thousands of random configurations of this knot type.

The writhe quantization of knots got even more puzzling when the comparison of writhe of ideal knots corresponding to all 85 prime knots with up to nine crossings revealed that their writhe values occupy only nine well-defined "levels" [6]. Analyzing writhe of ideal knots in the context of minimal diagrams of the corresponding knots, it was observed that the 3D writhe was an arithmetic sum of specific writhe values attributed to right- and left-handed torus and twist type of crossings in the minimal diagrams of alternating knots [7]. In 2000, we have demonstrated [8] that $\mathrm{W}r_{\mathrm{ideal}}$ of ideal knots and links (i.e., generalization of knots with several closed curves) can be predicted using an invariant of oriented alternating knots and links, namely the predicted writhe $P\mathrm{W}r$, which is a linear combination of the nullification writhe w_x and the remaining writhe w_y:

$$P\mathrm{W}r = \frac{10}{7}w_x + \frac{4}{7}w_y. \tag{6.1}$$

The nullification writhe w_x and the remaining writhe w_y are defined as follows [9]: transform a standard projection by nullifying (or smoothing) successive crossings until the unknot is reached, while, at each step, preventing the diagram from becoming disconnected. Then w_x is the sum of the signs of the nullified crossings and w_y is the sum of the signs of the remaining crossings. Examples of nullifications are shown in Figs. 6.2–6.4. A discussion about the coefficients 10/7 and 4/7 can be found in [8] and [5]. Since w_x and w_y are topological invariants, so is PWr, i.e., it depends on the topological type of the knot but not on a particular configuration of it.

The matching between PWr and Wr_{ideal} is strikingly good [8, 10]. This supports the notion that the ideal configuration contains important information about knots. Moreover, the fact that the calculation of 3D writhe of ideal knots can be performed using the minimal planar diagram of the knot greatly facilitates the calculation of the time-averaged writhe of randomly fluctuating knotted polymers. Complex simulations of ideal configurations are not needed but just a simple sort of crossings scoring in any minimal diagram of the corresponding knot. The method of writhe prediction that we proposed in [8] can be applied to any minimal diagram of alternating knots and links.

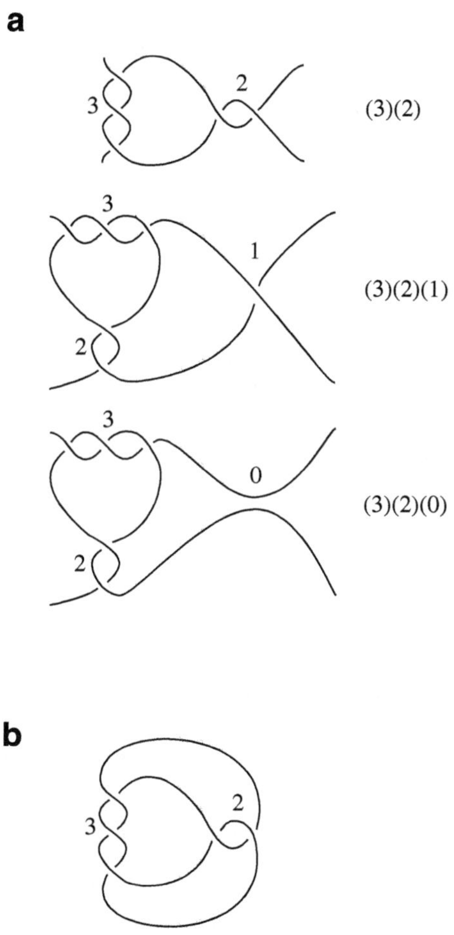

Fig. 6.1. (**a**) Examples of rational tangles. (**b**) The closure of a rational tangle gives rise to a rational link

If one considers all alternating knots and links, there is no simple function that can relate the writhe to the minimal crossing number. However, Huang and Lai [11] used (6.1) to calculate the writhe of ideal knots belonging to selected families of Conway knots, and observed that within these families the writhe can be expressed as a linear function of the minimal crossing number. We show here how to express the writhe as a linear function of the minimal crossing number in any individual Conway family of rational knots and link (Fig. 6.1). A related publication has been submitted to the *New Journal of Physics*.

6.2 Rational Tangles and Rational Links

Rational tangles and rational links have been introduced by Conway in 1970 [12]. A *rational tangle* is a region of a knot or link projection composed of a succession of vertical and horizontal rows of crossings, and is denoted by a sequence of numbers corresponding to the number of crossings in each row (see Fig. 6.2a). To avoid confusion, one always has to end with a horizontal row. If the latter contains no crossing, the sequence will end with (0). All crossings are done in order for the projection to be alternating (each strand alternatively goes over and under other strands). Figure 6.2a a shows positive rational tangles. If each crossing is inversed (i.e., the mirror image is considered), one gets negative rational tangles that will be denoted by a sequence of negative numbers. Conway proved that a rational tangle denoted by a mixed sequence of positive and negative numbers (i.e., a nonalternating projection) is always topologically equivalent to an alternating projection with all positive or all negative numbers. We will go on with positive rational tangles only. Extension to negative rational tangles is obvious.

The *closure* of a rational tangle is the operation of rejoining the two upper free ends and the two lower free ends on the projection (see Fig. 6.2b). The link we obtain is called a *rational link* (also called *two-bridge link* or *four-plat*). Rational links are completely classified. They all have either one or two components (let us recall that a one-component link is a knot). Nearly all knots and links naturally occurring in closed polymer chains are rational links.

6.3 Writhe of Families of Rational Links

We now examine some families of rational links and calculate their $P\mathrm{W}r$.

6.3.1 Tangles with One Row, Denoted by (a), a Positive Integer

Figure 6.3a shows an example of such a tangle, with $a = 5$. Since there is only one row, the minimum crossing number $n = a$ (5 in this case).

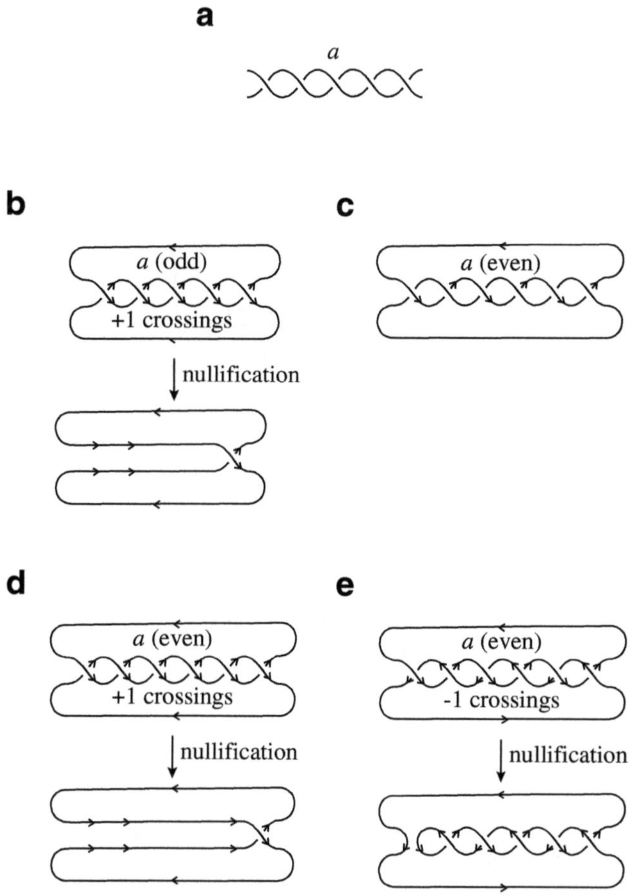

Fig. 6.2. (**a**) A rational tangle with one row of a crossings. (**b**) If a is odd, the closure of the tangle gives rise to a knot, whose nullification is shown. (**c**) If a is even, the closure of the tangle gives rise to a two-component link. Depending on the orientation chosen for the second component, two different situations occur, shown in (**d**) and (**e**)

a Odd

The rational knot obtained by the closure of such a tangle is the family containing knots $3_1, 5_1, 7_1$, etc. Figure. 6.3b shows the nullification process [9] applied to those knots. Crossings are successively nullified (or smoothed) until the unknot is reached, forbidding at each step the apparition of a disconnected component. The sum of the signs of the nullified crossings is w_x, the sum of the signs of the remaining crossings is w_y. In this case, nullifying $a-1$ positive

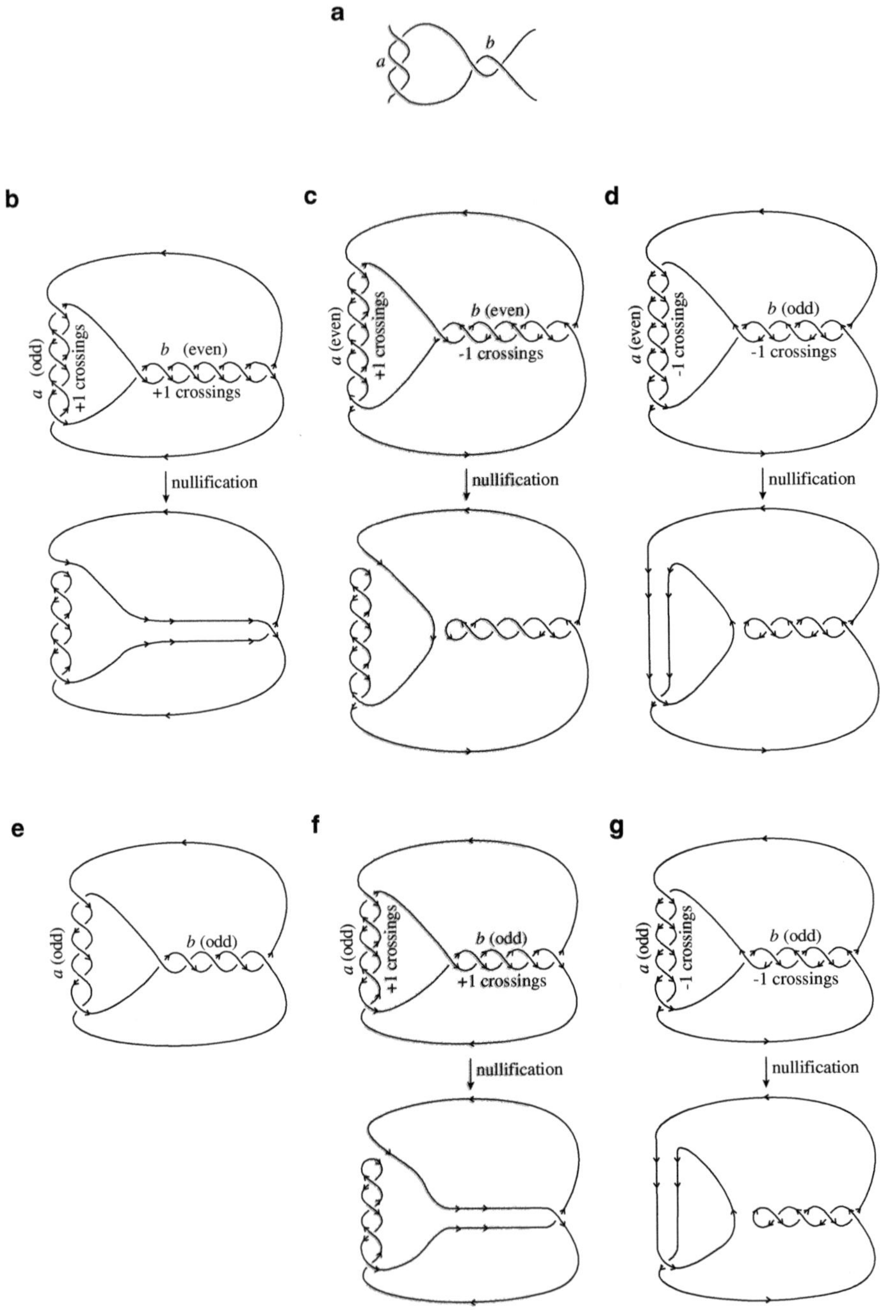

Fig. 6.3. (**a**) A rational tangle with two rows, containing a and b crossings, respectively. Drawings (**b**), (**c**), and (**d**) show the nullification of the knot obtained by the closure of the tangle in the cases a odd and b even, a and b even, a even and b odd, respectively. (**e**) In the case a and b odd, the closure of the tangle gives rise to a two-component link. Depending on the orientation chosen for the second component, two different situations occur, shown in (**f**) and (**g**)

crossings gives rise to the unknot. One cannot nullify the last crossing without disconnecting the link. So $w_x = a-1 = n-1$ and $w_y = 1$. Using (6.1) we get:

$$
\begin{aligned}
PWr &= \frac{10}{7}(n-1) + \frac{4}{7} \\
&= \frac{10}{7}n - \frac{6}{7}. \qquad (6.2)
\end{aligned}
$$

a Even

Figure 6.3c shows that the closure of such a tangle gives rise to a two-component link. There are therefore two possible orientations for the second component.

(a) Family of links $4_1^2 + -, 6_1^2 + +, 8_1^2 + +$, etc. (The symbols + and − refer to orientation. The convention used can be found in [13].) Figure 6.3d shows the nullification process applied to those links. Here again, we get $w_x = a-1 = n-1$ and $w_y = 1$. Therefore

$$
\begin{aligned}
PWr &= \frac{10}{7}(n-1) + \frac{4}{7} \\
&= \frac{10}{7}n - \frac{6}{7}. \qquad (6.3)
\end{aligned}
$$

(b) Family of links $4_1^2 + +, 6_1^2 + -, 8_1^2 + -$, etc. Figure 6.3c shows that the a crossings are now of negative sign and that only one crossing may be nullified. We already reach the unknot, and any further nullification would create a disconnected component. Thus $w_x = -1$ and $w_y = -(a-1) = -(n-1)$. That is to say, we have nullified one negative crossing and there remains $n-1$ negative crossings. Using (6.1) again, we get:

$$
\begin{aligned}
PWr &= -\frac{10}{7} - \frac{4}{7}(n-1) \\
&= -\frac{4}{7}n - \frac{6}{7}. \qquad (6.4)
\end{aligned}
$$

6.3.2 Tangles with Two Rows, Denoted by $(a)(b)$, a and b Positive Integers

Now, $n = a + b$. Figure 6.3a shows an example of such a tangle, with $a = 3$ and $b = 2$.

a Odd, *b* Even

The closure of this tangle gives rise to a family of knots with n positive crossings: a in the vertical row and b in the horizontal row. Looking at Fig. 6.3b, we see that we may nullify one positive crossing from the vertical row and $b-1$ positive crossings from the horizontal row. Thus $w_x = 1 + (b-1) = b$ and $w_y = (a-1) + 1 = a$. Using (6.1) and the fact that $a + b = n$, we have:

$$\begin{aligned} PWr &= \frac{10}{7}b + \frac{4}{7}a \\ &= \frac{4}{7}(a+b) + \frac{6}{7}b \\ &= \frac{4}{7}n + \frac{6}{7}b. \end{aligned} \tag{6.5}$$

Notice that when $b = 2$, we get the family of odd twist knots $(5_2, 7_2, 9_2, \ldots)$ and in that case

$$PWr = \frac{4}{7}n + \frac{12}{7}. \tag{6.6}$$

a Even, *b* Even

The closure of this tangle gives rise to a family of knots with a positive crossings in the vertical row and b negative crossings in the horizontal row. Figure 6.3c shows that we may nullify only one positive crossing from the vertical row and one negative crossing from the horizontal row. Thus $w_x = 1 - 1 = 0$ and $w_y = (a-1) - (b-1) = a - b$. (6.1) gives:

$$PWr = \frac{4}{7}(a-b). \tag{6.7}$$

Notice that when $b = 2$, we get the family of even twist knots $(4_1, 6_1, 8_1, \ldots)$. In that case, $a = n - 2$ and

$$\begin{aligned} PWr &= \frac{4}{7}((n-2) - 2) \\ &= \frac{4}{7}n - \frac{16}{7}. \end{aligned} \tag{6.8}$$

a Even, *b* odd

The closure of this tangle gives rise to a family of knots with n negative crossings: a in the vertical row and b in the horizontal row. Looking at Fig. 6.3d, we see that we may nullify $a-1$ negative crossings from the vertical row and

1 negative crossing from the horizontal row. Thus $w_x = -(a-1) - 1 = -a$ and $w_y = -1-(b-1) = -b$. Using (6.1) and the fact that $a+b=n$, we have:

$$\begin{aligned} PWr &= -\frac{10}{7}a - \frac{4}{7}b \\ &= -\frac{4}{7}(a+b) - \frac{6}{7}a \\ &= -\frac{4}{7}n - \frac{6}{7}a. \end{aligned} \tag{6.9}$$

a Odd, *b* Odd

Figure 6.3e shows that the closure of such a tangle gives rise to a two-component link. There are therefore two possible orientations for the second component.

(a) With the first possible orientation we get links with n positive crossings: a in the vertical row and b in the horizontal row. Looking at Fig. 6.3f, we see that we may nullify 1 positive crossing from the vertical row and $b-1$ positive crossings from the horizontal row, like in the case where a is odd and b even (case B.1). Thus $w_x = 1+(b-1) = b$ and $w_y = (a-1)+1 = a$. Using (6.1) and the fact that $a+b=n$, we have, as in case B.1,

$$\begin{aligned} PWr &= \frac{10}{7}b + \frac{4}{7}a \\ &= \frac{4}{7}(a+b) + \frac{6}{7}b \\ &= \frac{4}{7}n + \frac{6}{7}b. \end{aligned} \tag{6.10}$$

Notice that (6.5) and (6.10) are the same, but the first one (case B.1) deals with knots while we are now dealing with two-component links.

(b) With the other orientation for the second component, we get links with n negative crossings: a in the vertical row and b in the horizontal row. Figure 6.3g shows that we may nullify $a-1$ negative crossings from the vertical row and 1 negative crossing from the horizontal row, like in the case where a is even and b odd (case B.3). Thus $w_x = -(a-1) - 1 = -a$ and $w_y = -1-(b-1) = -b$. Using (6.1) and the fact that $a+b=n$, we have, as in case B.3,

$$\begin{aligned} PWr &= -\frac{10}{7}a - \frac{4}{7}b \\ &= -\frac{4}{7}(a+b) - \frac{6}{7}a \\ &= -\frac{4}{7}n - \frac{6}{7}a. \end{aligned} \tag{6.11}$$

The same remark holds, i.e., (6.9) and (6.11) are identical but the first one (case B.3) refers to knots while the second one refers to two-component links.

6.3.3 Tangles with Three Rows, Denoted by $(a)(b)(c)$, a, b, and c Positive Integers

Now, $n = a+b+c$. Figure 6.4a shows an example of such a tangle, with $a = 3$, $b = 1$ and $c = 2$. If we let a increase by steps of 2 and we fix b and c to 1 and 2, respectively, the closure of these tangles produces the family of knots $6_2, 8_2, 10_2$, etc. There are $2^3 = 8$ cases to study, depending on the parity of a, b and c. Let us illustrate the process with a odd, b odd, c even. There are a positive crossings in the first horizontal row, b positive crossings in the vertical row, and c negative crossings in the last horizontal row. Looking at Fig. 6.4b, we see that we may nullify $a-1$ positive crossings from the first row, 1 positive crossing from the second row, and 1 negative crossing from the last row. Thus $w_x = (a-1)+1-1 = a-1$ and $w_y = 1+(b-1)-(c-1) = 1+b-c$. Formula (6.1) gives:

$$PWr = \frac{10}{7}(a-1) + \frac{4}{7}(1+b-c). \tag{6.12}$$

Since $n = a+b+c$, if $b = 1$ and $c = 2$, then $a = n-3$ and we get for the family of knots $6_2, 8_2, 10_2$, etc.:

$$\begin{aligned} PWr &= \frac{10}{7}(n-3-1) + \frac{4}{7}(1+1-2) \\ &= \frac{10}{7}n - \frac{40}{7}. \end{aligned} \tag{6.13}$$

6.3.4 Tangles with r Rows

We can generalize this approach to any family of rational knots or links. Formula (6.1) will still hold, where each of w_x and w_y is a sum of the following form:

$$w_{x/y} = \underbrace{\left\{\begin{matrix} a-1 \\ -(a-1) \\ 1 \\ -1 \end{matrix}\right. + \left\{\begin{matrix} b-1 \\ -(b-1) \\ 1 \\ -1 \end{matrix}\right. + \left\{\begin{matrix} c-1 \\ -(c-1) \\ 1 \\ -1 \end{matrix}\right. + \cdots}_{r \text{ terms}} \tag{6.14}$$

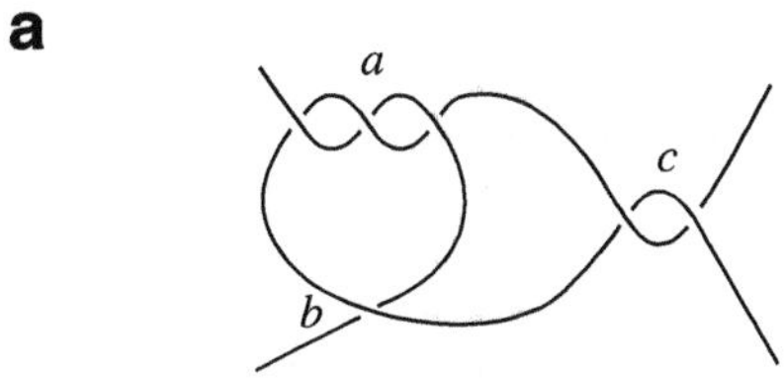

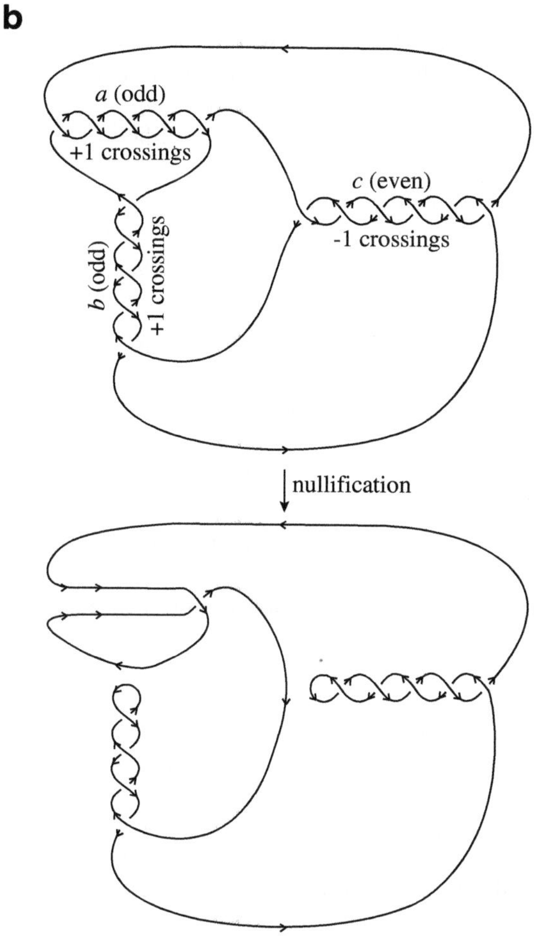

Fig. 6.4. (**a**) A rational tangle with three rows, containing a, b, and c crossings respectively. (**b**) Nullification of the knot obtained by the closure of the tangle in the case a odd, b odd, and c even

6.4 Discussion

6.4.1 When is PWr a Linear Function of n?

The simplest case is a family of rational tangles $(a)(b)(c)\ldots$ where all $a, b, c, \ldots$ are fixed except one. The nonfixed number is equal to n minus a constant since the sum of all $a, b, c, \ldots$ is n. Therefore in all those cases PWr is a linear function of n with slope $\pm 4/7$ or $\pm 10/7$.

We can then consider families of rational tangles $(a)(b)(c)\ldots$ where several of a, $b, c, \ldots$ change in a coordinated fashion, such that PWr is still a linear function of n. Two interesting cases in this regard are slopes ± 1 and 0. Let us first point out that when link orientation is indicated on rows of crossings, two situations occur. Either the arrows are antiparallel (crossings are called of *twist* type) and only one crossing will be nullified (e.g., row a on Fig. 6.3b), or the arrows are parallel (crossings are called of *torus* type) and all but one crossings will be nullified (e.g., row b on Fig. 6.3b). Each row, in turn, may be composed of positive or negative crossings. It follows from Formula (6.1) that a row with x crossings of twist type will contribute

$$\pm\left(\frac{10}{7} + \frac{4}{7}(x-1)\right) = \pm\left(1 + \frac{4}{7}x\right)$$

to PWr while a row with x crossings of torus type will contribute

$$\pm\left(\frac{10}{7}(x-1) + \frac{4}{7}\right) = \pm\left(\frac{10}{7}x - 1\right)$$

to PWr. If we consider a family of rational links where each successive member has two more positive crossings of twist type in a given *row* and two more positive crossings of torus type in another row, the increase in PWr will be of

$$\frac{4}{7} \cdot 2 + \frac{10}{7} \cdot 2 = 4$$

for an increase in n of 4 (four more crossings), leading to a slope of $+1$. As an example, the family composed of knots with tangles $(3)(2), (5)(4), (7)(6), \ldots$ have

$$PWr = 4 + \frac{4}{7}, 8 + \frac{4}{7}, 12 + \frac{4}{7}, \ldots$$

respectively. Similarly, if we consider a family of rational links where each successive member has two more negative crossings of twist type in a given row and two more negative crossings of torus type in another row, PWr will decrease:

$$-\frac{4}{7} \cdot 2 - \frac{10}{7} \cdot 2 = -4,$$

while n will increase by 4 (four more crossings), leading to a slope of -1.

Let us now examine a family of rational links where each successive member has two more positive crossings of a given type (twist or torus) and two more negative crossings of the same type. The predicted writhe is unchanged, leading to a slope of 0 versus n. An example is given by the family composed of knots with tangles $(3)(1)(2), (3)(3)(4), (3)(5)(6), \ldots$ whose members all have $P\text{Wr} = 20/7$. Notice that we do not examine cases of coordinated changes in rows by steps of one crossing at a time, because these steps may convert a row of crossings of twist type into a row of crossings of torus type or vice versa.

6.4.2 *P*W*r* of Achiral Knots

Since the 3D writhe is a measure of chirality of oriented closed curves in 3D space, it is a good test to see how achiral knots behave when seen as members of Conway families. The case of 4_1 is interesting. It belongs to the family of even twist knots $(4_1, 6_1, 8_1, \ldots)$ for which we have seen above (6.8) that PWr is a linear function of n with slope 4/7:

$$P\text{W}r = \frac{4}{7}n - \frac{16}{7}.$$

Let us replace n by 4 and, as by a miracle, $P\text{W}r$ becomes zero! Now, let us consider knot 4_1 as a member of another Conway family. 4_1 has rational tangle (2)(2) and 8_3, another achiral knot, has rational tangle (4)(4). We should thus be able to express that they belong to a Conway family with slope 0 versus n (since both $P\text{W}r$ must be equal to zero). Indeed, both considered tangles have two rows, one with positive crossings of twist type and one with negative crossings of twist type (see Fig. 6.3c) so adding two crossings to each row gives a net result of zero. We are in the case of a coordinated change in several rows leading to a slope of 0 versus n. The same Conway family contains knots with tangles $(6)(6), (8)(8)$, etc. all of which are achiral.

6.4.3 Shifts Between *P*W*r* as Linear Functions of *n*

Let us consider family $(a)(b)$, n odd, with b even and fixed (thus $a = n - b$ is odd) so by (6.5):

$$P\text{W}r = \frac{4}{7}n + \frac{6}{7}b$$

and compare it to family $(a)(b)$, n even, with b even and fixed to the same value (thus $a = n - b$ is even) so by (6.7):

$$\begin{aligned} P\text{W}r &= \frac{4}{7}(a - b) \\ &= \frac{4}{7}(n - b - b) \\ &= \frac{4}{7}n - \frac{8}{7}b. \end{aligned}$$

The two linear functions of n are thus shifted by $2b$. Since b is an even integer, the shift is a multiple of 4. To illustrate this, the linear function corresponding to the family of odd twist knots $(5_2, 7_2, 9_2, \ldots)$ with tangle $(n-2)(2)$, n odd, and the linear function corresponding to the family of even twist knots $(4_1, 6_1, 8_1, \ldots)$ with tangle $(n-2)(2)$, n even, are shifted by $2 \times 2 = 4$. It is worth noticing that the abscissas of the points on both lines are different: odd n's for the first line, even n's for the second line.

We may similarly compute the shift between any two linear functions of n having the same slope, and this can happen even if the families do not have the same number of rows in their corresponding tangles. For example, let us consider the family of torus knots with tangle (a), $a = n$ odd (not fixed), so by (6.2):

$$PWr = \frac{10}{7}n - \frac{6}{7},$$

and compare it to family $(a)(b)(c)$, n even, with b odd and fixed, c even and fixed (thus $a = n - b - c$ is odd) so by (6.12):

$$\begin{aligned} PWr &= \frac{10}{7}(a-1) + \frac{4}{7}(1+b-c) \\ &= \frac{10}{7}(n-b-c-1) + \frac{4}{7}(1+b-c) \\ &= \frac{10}{7}n - \frac{6}{7}b - 2c - \frac{6}{7}. \end{aligned}$$

The shift is of $6/7b + 2c$. To take a specific case, if we compare the family of torus knots to the family of knots $6_2, 8_2, 10_2, \ldots$ having tangle $(n-3)(1)(2)$, (6.13) gives:

$$PWr = \frac{10}{7}n - \frac{40}{7},$$

and the shift is of $10/7 + 4 = 34/7$.

6.4.4 Knots Versus Two-Component Links

As already mentioned, the closure of a rational tangle gives rise either to a knot or to a two-component link. Actually, this can be directly related to the parity of the nullification writhe w_x (see Prop. 12 of [9]). For a family of knots w_x is even while for a family of two-component links w_x is odd.

6.5 Conclusion

Using the formula $PWr = 10/7w_x + 4/7w_y$ introduced in [8], we can predict the 3D writhe of any rational knot or link in its ideal configuration, or equivalently, the ensemble average of the 3D writhe of random configurations of it. For every family of knots or links corresponding to a rational tangle $(a)(b)(c)\dots(r)$ having a fixed number of crossings on $r-1$ rows, PWr presents a linear behavior versus n, the minimal crossing number of the knot or link, with a slope of $\pm 4/7$ or $\pm 10/7$. One can also consider families of rational tangles $(a)(b)(c)\dots(r)$, where several of $a, b, c, \dots$ change in a coordinated fashion, such that PWr is still a linear function of n. It is also possible to compute the shift between two lines having the same slope. We thus have at our disposal a formalism allowing to predict a number of data usually obtained numerically, and which might help to shed some light on the "quantum mystery of knots" [1].

Acknowledgments

C. C. is a Research Associate of the Belgian FNRS (National Fund for Scientific Research). This work was supported in part by Swiss National Science Foundation Grant 31-61636.00 to A. S.

References

1. K. Devlin, New Scientist, 40–42 (2001)
2. V. Katritch, J. Bednar, D. Michoud, R.G. Scharein, J. Dubochet, A. Stasiak, Nature **384** 142–145 (1996)
3. E.J. Janse van Rensburg, E. Orlandini, D.W. Sumners, M.C. Tesi, S.G. Whittington, Phys. A: Math. Gen. **26** L981–L98 (1993)
4. E.J. Janse van Rensburg, D.W. Sumners, S.G. Whitington, in *Ideal Knots*, ed. by A. Stasiak, V. Katritch, L.H. Kauffman, (World Scientific, Singapore, 1998), pp. 70–87
5. A. Stasiak, J. Dubochet, V. Katritch, P. Pieranski, in *Ideal Knots*, ed. by A. Stasiak, V. Katritch, L.H. Kauffman (World Scientific, Singapore, 1998), pp. 1–19
6. P. Pieranski, in *Ideal Knots*, A. Stasiak, V. Katritch, L.H. Kauffman (World Scientific, Singapore, 1998), pp. 20–41
7. A. Stasiak, in *Proceedings of the Delphi Conference on Knots*, ed. by C. Gordon, V.F.R. Jones, L. Kauffman, S. Lambropoulou, J.H. Przytycki (World Scientific, Singapore, 2000), pp. 477–500
8. C. Cerf, A. Stasiak, Proc. Natl Acad. Sci. USA 97 3795–3798 (2000)
9. C. Cerf, J. Knot Theory Ramif. **6** 621–632 (1997)
10. P. Pieranski, S. Przybyl, Eur. Phys. J. E **6** 117 (2001)
11. J.-Y. Huang, and P.-Y. Lai, Phys. Rev. E **63** 021506 (2001)
12. J.H. Conway, in *Computational Problems in Abstract Algebra* (Pergamon Press, Oxford, 1970), pp. 329–358
13. C. Cerf, Topology Atlas Invited Contributions **3** 1–32 (1998). http://at.yorku.ca/t/a/i/c/31.htm

7

Combinatories and Topology of the β-Sandwich and β-Barrel Proteins

A.E. Kister, M.V. Kleyzit, T.I. Gelfand, and I.M. Gelfand

Summary. One of the main challenges in life science today is to understand how genomic sequences determine geometric structure of proteins. Knowledge of the three-dimensional structure provides valuable insights into functional properties of proteins, since function of proteins is largely determined by their structure. The ability to classify a genomic or amino acid sequence into its proper protein family, and thereby to predict, to some degree of approximation, its structure and function is an essential prerequisite to using genomic information for explaining enzymatic processes that underlie cell behavior, understanding the molecular basis of disease, and achieving rational drug design.

7.1 Introduction

With more than fifty complete genomes already sequenced, and at least a hundred more close to completion [1], the gap between known sequences and solved structures (collected at the Protein Data Bank [2] and classified in the SCOP database [3]) is quickly widening. Consequently, the task of structure prediction from amino acid sequence has taken center stage in the "postgenomic" era.

Direct approaches to structure determination include X-ray crystallography, and nuclear magnetic resonance, among other techniques. However, such methods are expensive, time consuming, and not always applicable.

The potential of alternative methods for protein comparison and classification is not settled yet, and there is an urgent need for more reliable approaches for such bioinformatics problems. Alternative approaches based on theoretical study of the nature of the sequence/structure relationship can be immensely useful in dealing with a wealth of data on newly sequenced genomic sequences.

Although it is more than 40 years since we know that all information required for the folding of a protein chain is contained in its amino acid sequence, we have not yet learned how "to read" this text as to predict the detailed 3D structure a protein whose sequence is known [4].

There is both a local and a global point of view regarding the relationship between the linear sequence of amino acids and the resulting three-dimensional structure of protein. The former viewpoint postulates just a few critical residues, some 10–20%, of the sequence play the most critical role in determining the characteristics of a fold, while the latter considers all residues in the sequence as crucial [5, 6].

The most commonly used methods of the global sequence comparison (BLAST and FASTA [7–9]) match new sequences (queries) against all the sequences in a database (target) and report each query-target pair that represents a statistically significant match. At present, some of the most powerful approaches for protein classification are based on hidden Markov model [4–6]. However, it was shown that, as the sequence identities of related proteins go below 30% identity, the chance of their relationship being detected these methods becomes increasingly small. Thus, there is no doubt that the methods described above have been very successful for protein classification; however on the other hand, they all become less reliable as more distant, less homologous proteins are considered.

The local model received considerable support when Chothia and Lesk showed, that rather different amino acids sequences share the same fold, i.e., same major secondary structure in the same arrangement and with same chain topology [10]. In our recent study with Chothia and Lesk, we discussed why structure changes slower than sequence in protein evolution [11]. For related proteins, structure similarities arise in the course of their evolution from a common ancestor, while for proteins with very low homology fold similarity may be owed to physical and chemical factors. That favor certain arrangements for secondary structure units and chain topology.

The considerable step ahead in our understanding of how the amino acid sequence of proteins dictates its three-dimensional structure is a division of amino acids in the sequence into hydrophobic interior and a surface of a protein that is sufficiently hydrophilic. In our work we showed that residues of the hydrophobic interior make the major contribution to the stability of a protein [12]. Following George Orwell (*Animal Farm*), it can be concluded that not all residues are equally significant in how they contribute to the protein folding. Thus, the search of the key, conserved residue, i.e., residues that are "more equal" than other residues in a protein, is the essential step in solving the problem of the relationship between an amino acid sequence and a geometric structure of proteins.

In this work, we suggest a new method of protein classification based on the ideas of the local model. The main novelty of the method is in the identification of the key residues or sequence determinants. The sequence determinants serve as a basis for development of computer algorithms for protein classification and structure/function prediction of genomic and amino acid sequences. A direct corollary of the approach is that the complexity of protein sequence search algorithms and 3D structure predictions can be dramatically reduced.

The search is carried out with predefined sets of several (8–12) sequence determinants, instead of analysis of a whole protein sequences.

We focus in this work on the analysis of two large groups of the β-proteins sandwich-like and the barrel-like proteins. The goal of this research is to define the structural and sequence features, which these very different proteins have in common. Our first task is to analyze the supersecondary substructure of proteins to determine whether they have features that are invariant. Another aspect of our research involves finding conserved positions in sequences that are occupied by similar residues in all proteins.

Analysis of the supersecondary structure is based on the information about hydrogen bonds contacts between the main chain atoms of residues. It presents as a list of number of residues, which are connected by the hydrogen bonds between the main chain atoms of residues. It turned out that examination of only the list of the numbers of the hydrogen bonded residues was sufficient to determine: (a) a secondary structure; (b) an arrangement of strands (a supersecondary structure); (c) a protein fold (structural classification of the protein); (d) a set of the rules that governs the arrangement of the strands in the barrel and sandwich structures; (e) the supersecondary patterns: two pairs of strands, whose location in the structure is common in all barrel and sandwich proteins. This analysis discovered that despite a seemingly unlimited number of arrangements of strands there exists a rigorously defined constraint on supersecondary structures.

Another aspect of our research involves finding positions in sequences that are occupied by similar residues. The problem of the discovery of conserved positions is not easy for highly various groups of proteins as sandwich and barrel proteins. A comparison of the amino acid sequences in various superfamilies showed that the sequences are so diverse that even the most powerful approaches such as PSI-BLAST and hidden Markov model cannot find any sequence homology. However, the delineation of the invariant supersecondary, substructure, common for different superfamilies and protein folds, makes possible a secondary structure-based multialignment. It results in the set of the key conserved positions, whose residues share both structural and chemical properties and have a decisive role in geometry of the beta proteins.

Thus putting the information about amino acid sequences and hydrogen bond contacts together we give in account the principal relations between sequence and structure in the beta proteins.

7.2 Overview of the Structures

All β-proteins can be divided into several groups in relation to their "structural design" like β-sandwiches, β-barrels, β-propellers, and others [13]. Each of these architectural designs encompasses a large number of protein families, superfamilies, and folds within the structural hierarchical classification adopted in the SCOP and CATH databases [3, 14]. Proteins grouped together based on their common architectural properties often do not have share any functional homology or significant sequence homology.

Systematic theoretical analysis has revealed certain structural features common for the β-sheet as for example, Greek key and jellyrolls patterns and specific characteristics of the edge strands [15, 16]. In many works the geometric parameters of the β-proteins such as distance or angle matrices between strands, the number of strands and shear number of proteins and the analysis of bifurcation of β-sheets, and the spatial organizations of secondary structures [17–27] are investigated.

Analysis of arrangement of strands of the sandwich and barrel proteins revealed that only a limited number of possible strand arrangements are realized in existing structures. We found that each fold structural classification of the barrel and the sandwich structures can be described by a unique arrangement of strands. This finding has implications for protein classification. Since strand arrangement of a protein can be deduced from its matrix of hydrogen bonds, it follows that one can in most cases assign a query protein to its proper fold given sufficient information about its hydrogen bonds.

7.3 Common Features in Structures and Sequences of Sandwich-Like Proteins

7.3.1 General Features of the Sandwich-Like Proteins

Proteins of 69 superfamilies in 38 protein folds have been described as "sandwich-like proteins" (SPs) (see folds 1.2.1 – 1.2.38 in SCOP (3), release 1.59). Spatial structures of SPs are composed of β-strands, which form two main β-sheets that pack face to face. Although the general architecture of SP is relatively uniform, the number of strands and the arrangement of the strands vary widely [28–30]. Some SPs, in addition to two "main" sandwich sheets, contain "auxiliary" beta sheets. Comparison of proteins in different superfamilies does not show either functional homology or significant sequence homology.

7.3.2 Supersecondary Patterns in the Sandwich-Like Proteins

The determination of H-bonds between the main chain atoms allowed us to determine the arrangements of the strands and identify those strands that make up the two main sandwich sheets. Analysis of the arrangements of strands in all known sandwich-like known protein structures revealed the definite rules that are valid for almost all sandwich proteins.

At first, this rule can be stated as follows: in any given sandwich-like protein structure there exist two pairs of strands $(i, i+1)$ and $(k, k+1)$, such that: (1) strands of each pair are adjacent to each other in sequence (Fig. 7.1a); (2) strand i is located in one main sheet and $i+1$ – in the other

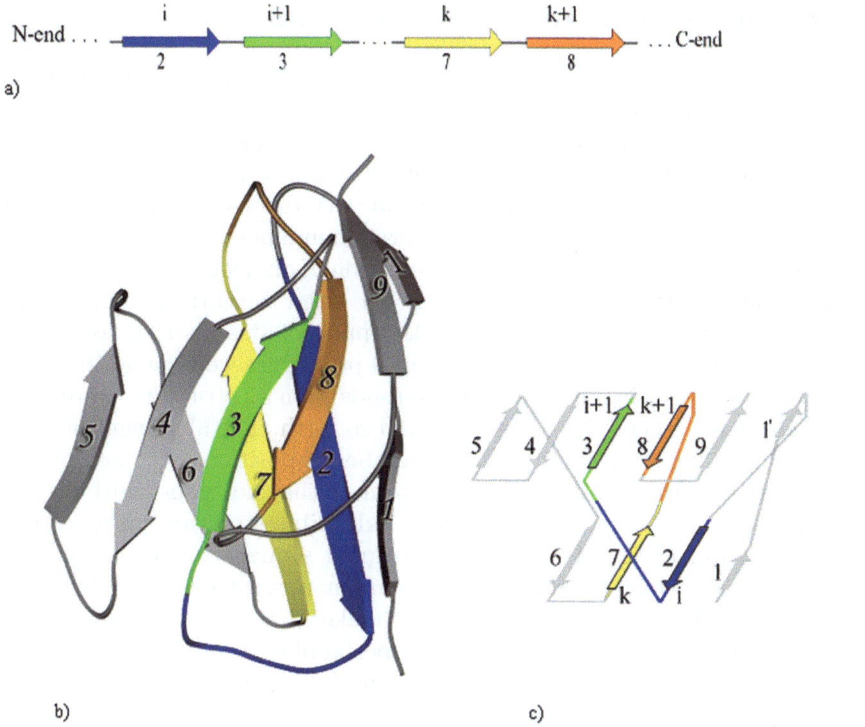

Fig. 7.1. The schematic representation of an immunoglobulin variable domain (sandwich-like family). (**a**) β-Sheet strands are numbered sequentially as they are presented in a sequence. The strands 2, 3, 7, and 8 are shown; (**b**) Chain fold of immunoglobulin variable domain of heavy chain (PDB code: 1ine). The drawing is done using the MOLSCRIPT program [31]. β-Sheet strands are shown as ribbons. (**c**) Arrangement of the strands in two main β-Sheets. The interlocked pairs of strands $(i, i+1)$ and $(k, k+1)$ correspond to the strands #2, 3 and 7, 8

main sheet; (3) strand k is found in one main sheet and $k+1$ – in the other; (4) strands i and k are located within the same sheet; they are antiparallel to each other and linked by hydrogen bonds; (5) likewise, strands $k+1$ and $i+1$ are located within the other main sheet, are antiparallel to each other, and H bonded. (Fig. 7.1b, c).

These two interlocked pairs form a sandwich-like substructure within SP. Usually they are found in the middle of the sheets. The number of strands interposed between strands i and k varied from 1 to 10, but in 80% of cases the number of interposed strands was 2 – 4 (see fig. 7.1c and 7.2b; where the number of strands between the strands i and k is equal to 4 and 2, respectively). Two interlocked pairs were detected in 94% of all analyzed structures. Thus, this investigation led to the discovery of a central feature of the super-secondary structure that is invariant in almost all SP.

7.3.3 Structurally Based Sequence Alignment

The essential element of our method is that it involves alignment not of whole sequences, but of corresponding strands in their respective proteins. In contrast to the analysis of homology proteins in protein family, the determination of the corresponding strands in a group of strongly varied proteins, such as a collection of sandwich-like proteins from very diverse superfamilies with a dissimilar arrangement of strands and variable numbers of strands, could be a complicated problem. However, the delineation of an invariant supersecondary substructure made it possible to identify and align the corresponding strands.

It follows from the rule of interlocked pairs of strands that four strands $i, i+1, k$, and $k+1$ with similar structural properties were found in all sandwich proteins. Thus in our procedure i strands from all structures were aligned with each other, then all $i+1$ strands and so forth. The alignment of strands carried out in this way maximizes the number of positions occupied by structurally similar residues. It is important to note that no "gaps" within strands are allowed, since strands are viewed as indivisible structure units. Adjacent residues within a strand are always assigned sequential position numbers. However, gaps between strands are a common occurrence. The advantage of this structurally based approach is that it makes possible a common system of numbering for sequences from different superfamilies. It allowed us to compare nonhomologous SP. Details of the structure-based sequence multialignment and a list of the conserved positions was presented in our works [32, 33].

7.3.4 Sequence Characteristics of the $i, i+1, k$, and $k+1$ Strands

Analysis of the structurally aligned sequences revealed 12 positions, which are occupied by residues with structurally similar properties in their respective SP structures. We suppose that residues that have the same structural properties across all SPS are their structural determinants. The structural determinants lie at the center of the interface between the β-sheets and form the common geometrical core of SP structures.

Inspection of amino acid frequencies in these 12 positions showed that eight positions are the conserved hydrophobic positions of SP. Residues at these eight positions are termed the SP sequence determinants. Eighty percent of all SP sequence determinants are *V, L, I, and F residues.*

7.3.5 Structural Features of the Sequence Determinants

Relative positions of the eight sequence determinants in $i, i+1, k$, and $k+1$ strands are common in all SP structures that contain interlocked pairs (Fig. 7.2). The strands i and k are oriented toward each other in such a way that residues at position 6 in strand i always form main chain hydrogen bonds with residues at position 8 in strand k, and side chains of both these residues look inside the hydrophobic interior of SP. Orientation of strands $i+1$ and

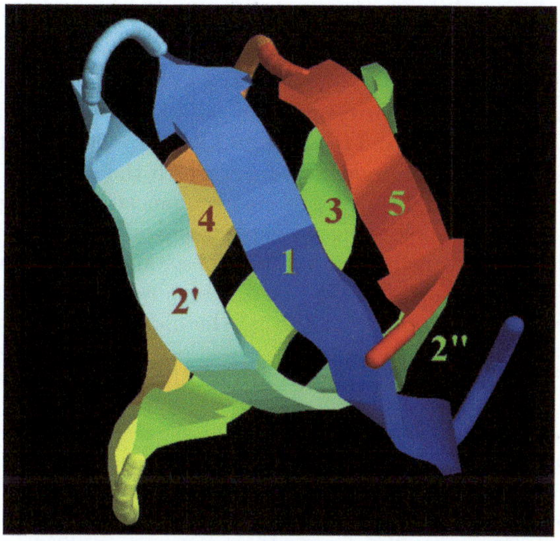

Fig. 7.2. Schematic representation of the barrel structure (1bia). The main β-sheet strands are numbered as they presented in the sequence. The strand 2 is divided into $2'$ and $2''$ parts. The long strand 2 is bent into two parts: $2'$ and $2''$. The $2'$ part forms H-bonds with strand 1 in one part of the β-sheet, while the $2''$ part has hydrogen contacts with strand 3 in the second part of the β-sheet. The strands 1, $2'$, and 5 form one subsheet and the strands $2''$, 3, and 4 form the second subsheet

$k+1$ is also fixed: residues at position 6 in $i+1$ and position 8 in $k+1$ strands are located opposite to each other in the sheet, but do not form H-bonds. Side chains of both residues look inside the hydrophobic interior of SP.

7.3.6 Method of the Sequence Determinants for Identification of Proteins

Knowledge of sequence determinants of protein groups allows us to develop a computer algorithm for classification of proteins. A set of sequence determinants is characterized by (1) a number of the sequence determinants for a given protein group (usually there are 8–12 sequence determinants); (2) a set of residues that characterize each sequence determinant, and (3) intervals (a number of residues) between the sequence determinants in sequences. This data will be used to distinguish all proteins of a particular protein group and to predict the secondary and tertiary structure.

For the search procedure, we implemented an algorithm based on appropriate modification of the dynamic programming [33]. This algorithm matches one-by-one sequence determinants of a given protein group with residues of the query sequence. Once a match is found for the sequence determinant closest to the beginning of the sequence, the algorithm starts to look for a match for the second determinant in the query sequence, and so on. If all sequence determinants are matched, then the protein is assigned to the group.

Table 7.1. Identifying four families of sandwich-like proteins within 11 distinct genomes four protein families are classified as in SCOP database: (1) (PL) – protein family of lipoxygenase N-terminal domain; (2) (AT) – protein family: Alpha-toxin, C-terminal domain; (3) (AD) corresponds to 30-kd adipocyte complement-related protein; (4) (TR) corresponds to TRANCE/RANKL cytokine protein domain. The first column lists the names of organisms from which the genomes are derived. The second column contains numbers of proteins sequenced from respective genomes. The number of sequences belonging to each group of proteins (PL, AT, AD, or TR) found in the genome using our method of sequence determinants (MSD) is given in the "MSD" columns. "HMM" columns show the number of sequences of the respective groups of proteins found using the hidden Markov models

Genomes	Proteins	HMM	MSD	HMM	MSD	HMM	MSD	HMM	MSD
Arabidopsis thaliana	25617	8	11	4	5	0	0	0	0
Clostridium acetobutylicum	3672	0	1	1	2	0	0	0	3
Clostridium perfringens	2660	0	2	1	1	0	0	0	0
Mesorhizobium loti	6752	1	1	0	2	0	0	0	0
Pseudomonas aeruginosa	5567	0	0	1	0	0	0	0	0
Caenorhabditis blegans	20448	5	9	0	0	0	0	0	0
Drosophila melanogaster	14335	2	5	0	0	0	0	1	1
Escherichia coli K12	4289	0	0	0	1	0	0	0	0
Escherichia coli 0157H7	5361	0	1	0	1	0	0	0	0
Bacillus halodurans	4066	0	0	0	0	0	0	0	0
Lactococcus lactis	2266	0	1	0	0	1	1	0	0

The results of applying the search algorithm that uses sequence determinants of 4 protein families in 11 different genomes are presented in Table 7.1. MSD' column of the table contains data on how many proteins of the given family were found in the respective genome through application of our algorithm. For comparison purposes, "HMM" column gives the number of proteins of the family found using HMM search procedure, considered to be the most powerful of currently used method [34].

Overall, both methods found approximately the same number of SPs in the 11 genomes. All sequences founded by HMM were detected by our approach (except one). However, our method revealed a number of additional sequences that can be putatively assigned to the four families. For the most part, these "additional" proteins are labeled as "unrecognized proteins" in the genome. It is suggested that our approach can identify even those SPs that are "hidden" from HMM search procedure. Further investigations are necessary to tell whether these "candidate" sequences indeed qualify to join the respective SP families. Our approach also provides an independent check on the accuracy of HMM-based algorithm.

7.4 Common Structural and Sequence Features of Barrel-Like Proteins

7.4.1 Search for Sequence and Structural Invariants in Barrel Proteins: An Outline of the Approach

In this work we analyze two barrel characteristics crucial for sequence/ structure relationship, which are not still consider in detail. They are the arrangement of strands in the β-sheet and a characteristic of a "place of a distortion" in the β-sheet, i.e., a place where strands curve to form a barrel structure. These features distinguish "one β-sheet geometry" of the barrel proteins.

Our approach is based on a comparison of elements of secondary structures with analogous structural role in their respective barrel proteins. The first step is to define precisely the strands that make up the main sheet and determine the arrangements of the strands in space. Special attention is devoted to the key region of the barrel proteins, the right-angle "turn" in supersecondary structure. Examination of the strands that make up this crucial region allows us to identify invariant substructure present in all barrel structures. Analysis of amino acid sequences of the substructure led to discovery of conserved positions: sequence invariants of the barrel proteins.

7.4.2 Overview of the β-Barrel Structures

In general, the main β-sheet of the barrel proteins is folded to form a closed structure. However, some proteins, though structurally very similar to "closed barrels" and possessing a characteristically bent main sheet, have their edge strands too far from each other in space to form hydrogen bonds. These proteins are referred to as "partly open barrels" in SCOP database nomenclature and are grouped together with closed barrels [2]. Each of the strands of the main β-sheet forms hydrogen bonds with two or three adjacent strands.

McLachlan [9] has shown that the main structural parameters of the barrel structures are the number of strands that form the β-sheet, n, and the measure of its stagger, "the shear number," S [17]. Later it was shown that these parameters define the geometry of barrel structures ([19, 20], see, as well, [27]).

According to the SCOP database, release 1.61 [2], one or more domains in 2311 protein structures form barrel-like structure. For example, RNA-binding (SM-like) protein, listed as structure 1i4k in PDB, contains 28 barrel-like domains. Barrel-like structures make up 114 protein families, 66 superfamilies, and 36 protein folds (##40–75 of "All beta" proteins in the SCOP database). Proteins in different barrel superfamilies do not share either functional homology or sequence similarity.

7.4.3 Defining of the β-Strands and Loops

As mentioned above our Structural analysis is based on the information about H-bonded residues. A pair of β-strands is determined as two fragments of residues whose residues are connected by H-bonds between the main chain atoms. The first residue in the segment to form an H-bond is considered to be the first residue of the strand, while the residues that come before it in a sequence are not involved in H-bonding and are, therefore, considered to be part of a loop. Similarly, the last residue of an amino acid segment to form an H-bond defines the end of the strand. Based on this analysis the strands in the structure are consequently numbered starting from N-end of sequences.

The segments of residues between the strands are considered as the loops. We do not consider here the conformation of loops. As known from the analysis of all beta proteins some loops in the structures were observed in the helical conformations.

Strands in barrel structures are numbered sequentially starting from the N-end of the sequence. Barrel structures differ by the number of strands ("n") that form the cylinder. In the "All beta" class of SCOP database beta-barrel structures are made from 4 to 8 strands.

Some strands in the main β-sheet have more than two adjacent strands. These are termed "bifurcated strands." More than half of the barrel structures have at least one such bifurcated strand. As a result, they contain the so-called "side" β-sheets in addition to the main barrel β-sheet. So far as our goal is to define what is common to all barrel structures, we eliminate the side β-sheets from our analysis of supersecondary structures, and retain only the strands that make up the main β-sheet. The strands in the "edited" barrel structures are numbered sequentially starting from the N-end of the sequence.

Barrel structures differ by the number of strands (n^*) that form cylinder structure. In the "All beta" class of SCOP database beta-barrel structures contain 4–8 such strands.

7.4.4 Arrangement of the Strands in the β-Sheet

The interstrand hydrogen bonds define the strand arrangement of β-sheet. The H-bonds were calculated for all barrel structures. Figure 7.2 illustrates the interconnection among strands in the 1bia structure: strand 1 is H-bonded to strands 2 and 5, while strand 5 is bonded only to strand 1; strand 2 is bound to strands 1 and 3, and strand 4 – to strand 3. Since strands 4 and 5 are not connected to each other, 1bia is a partly open barrel. Arrangement of strands of main β-sheet in the 1bia structure can be represented schematically as: 5-1-2-3-4. Arrangement of strands in the beta barrels of different folds is presented in the "Arrangement" column of Table 7.2.

Despite the large number of possible combination of strands in barrel structures, the number of different arrangements of strands is in reality very limited. For example, the theoretical number of strand combinations in the barrel

Table 7.2. The arrangement of the strands in the barrel structure Columns: F – Protein folds in the SCOP classification; Fm – the numbers of the given fold; arrangement – arrangement of the strands in the main beta sheet; subsheets – arrangement of the strands in two subsheets

Fold	Str	Arrangement							Sub-sheets							
40	1bia	5	**1**	**2**	**3**	**4**			B:			**2′**	**1**	5		
									A:		2″	**3**	**4**			
41	1jh2	**4**	**1**	**2**	**3**	5			B:		**2′**	**3**	5			
									A:	2″	**1**	**4**				
42	1pdr	**1**	**5**	**4**	**3**	2			B:			**4′**	**3**	2		
									A:		4″	**5**	**1**			
43	1g3p		**2**	**3**	**4**	**1**			B:	4″	**3′**	**2**				
									A:	3″	**4′**	**1**				
44	1b34	5	**1**	**2**	**3**	**4**			B:		2″	**3′**	**4″**			
									A:	4′	3″	**2′**	**1**	5		
45	1h5p	5	**1**	**2**	**3**	**4**			B:			**3**	**4**			
									A:			**2**	**1**	5		
46	1whi	**1**	**2**	**3**	**4**	5	1		B:		2″	**3′**	**4**	5		
									A:		3″	**2′**	**1**			
47	1sty	**1**	**2**	**3-**	-5	**4**	**1**		B:	1″	**2**	**3**	5′			
									A:		**1′**	**4**	5″			
48	1dxr	**1**	**2**	**3**	**4**	5	6-	-1	B:		2″	**3′**	**4**			
									A:		3″	**2′**	**1**	6	5	
49	1bfg	1	**2**	**3**	4	5	**6**	**1**	B:		**2**	**3**	4	5		
									A:		**1**	**6**				
50	1i8d	1	2	5	**4**	**3**	**6**	**1**	B:		6″	**3**	**4**	5	2	1″
									A:			**6′**	**1′**			
51	1efc	**1**	**4**	**3**	**2**	5	6	1	B:		**3′**	**2**	5	6″	1′	4″
									A:	3″	**4′**	**1″**	**6′**			
52	1flm	2	1	**4**	**5**	**6**	**3**		B:		6″	**5′**	**4**	1	2	
									A:		5″	**6′**	**3**			
54	1fmt	**1**	**2**	**5**	**4**	3	6		B:	2″	**5′**	4	3	6		
									A:	5″	**2′**	**1**				
55	1eax	**1**	**2**	**3**	**6**	5	4	1	B:			**2′**	**1**	4	5′	
									A:		2″	**3**	**6**	5″		
56	1bco	1	4	5	6	3	2		B:		**5′**	**6**	3	2		
									A:	5″	**4**	**1**				
57	1e79	1	**2**	**5**	**4**	**3**	6	1	B:			**4′**	**3**	6	1′	2″
									A:		4″	**5**	**2′**	1″		

Table 7.2. Continued.

Fold	Str	Arrangement	Sub-sheets								
58	1k8h	1 6 **7** **2** **5** **4** 3	B:		**5**	**4**	3				
			A:		**2**	**7**	6	1			
59	1ile	1 2- -6 3- -5 4 1	B:			**1′**	**2″-**	-6	3-	5″	4′
			A:	2′	1″	**4″**	**5′**				
60	1gmu	**2** **1** **4** **5** 3 6	B:		**4**	**5**	3	6			
			A:		**1**	**2**					
61	2eng	3 4- -1 5 6 2	B:		1″	**5**	**6-**	-2			
			A:			**1′-**	**-4**	3			
62	1dfu	1- -5 2- -4 3 6- -1	B:		**2′-**	**-4**	3	6			
			A:	2″	**5-**	**-1**					
63	1h9d	1 3 **2-** **-6** **5** **4-** -1	B:			**5′**	**4-**	-1			
			A:		5″	**6-**	**-2**	3			
64	1mai	3 2 **1** **6** **5** **4**	B:	6″	**5**	**4**					
			A:		**6’**	1	2	3			
65	1ytf	**1** **2** 3 4 **5** **6** 1	B:			**6**	**5**	4″			
			A:			**1**	**2**	3	4′		
66	1ieg	1 2 3	B:	1″	**2′**	**3″**					
			A:	3′	**2″**	**1′**					
67	1pkm	**1** **4** **3** **2-** -6 5 7 1	B:			**3′**	**2-**	-6	5		
			A:		3″	**4**	**1**				
68	1ik9	**1** **7** **4** **5** 6 2 3 1	B:		4	5	6	2			
			A:		**7**	**1**	3				
69	1hbq	1 2 **3** **4** **7** **6** 5 1	B:	1	2	**4**	**3**	2	1		
			A:			**7**	**6**	5			
70	3	1 2 **3** **4** **5** **6** 7 8 1	B:	5″	**4**	**3**	2	1′			
			A:		**5′**	**6**	7	8	1″		
71	1swu	1 2 **3** **4** **5** **6** 7 8 1	B:			**4′**	**3**	2	1		
			A:		4″	**5**	**6**	7	8		
72	2cpl	1 8 **1** **2** **7** **5** 6 4 3	B:	7″	**2**	**1**	8				
			A:		**7′**	**5**	6	4	3		
73	1ija	1 2 3 **4** **8** **7** **6** 5 1	B:			**7′**	**6**	5	1	2	
			A:		7″	**8**	**4**	3			
74	1c39	1 2 3 4 **7** **8** **6** **5** 1	B:		**6′**	**5**	1	2	3		
			A:	6″	8	7	4	3″			
75	1f3u	1 2 **3** **4** **5** **6** 8 7	B:			**4′**	3	2	1		
			A:		4″	**5**	**6-**	-8	7		

structures, which are formed by six strands ($n^* = 6$), is equal to 360 (6!/2). However, only eight variants with different arrangements are found among all beta-barrel structures in SCOP database. These eight variants can be represented as follows: (1) 1-2-3-4-5-6-1 (folds ## 48, 49,64, 65); (2) 1-2-5-4-3-6-1 (folds ## 50, 54, 57); (3) 1-4-3-2-5-6-1 ((folds ## 51, 55); (4) 1-2-3-6-5-4-1 (folds ## 52, 56); (5) 1-3-2-6-5-4-1 (fold # 63); (6) 1-2-6-3-5-4-1 (folds ## 59, 60); (7) 1-5-6-2-3-4-1 (fold # 61); (8) 1-5-2-4-3-6-1 (fold # 62).

Thus from our analysis follows that structures from different folds can have the similar arrangement of the strands (see variants ## 1, 2, 3, 4 and 6 above).

7.4.5 Two Subsheets in the Barrel Structures

The main β-sheet of barrel structures can be divided into two groups of strands or two subsheets. The strands in each subsheet are approximately parallel to each other, and make an angle of about 90° with the strands of the other subsheet. This orthogonal beta sheet packing makes possible the formation of the cylindrical "barrel" structure.

7.4.6 Four Types of Connection Between the Strands in Two Subsheets

There are four ways whereby strands of two orthogonal subsheets can come together to form a single β-sheet in the barrel structures (Fig. 7.3):

(a) The edge strands a and k are located in close proximity to each other allowing for one or two H-bonds to be formed between the two strands (Fig. 7.3a).
(b) Two orthogonal strands are connected by means of a long, 90° bent strand. Let us denote the two "legs" of the long strand as k′ and k″ (Fig. 7.3b). Residues of k′ part form H-bonds with residues from the strand in one subsheet (strand m), while residues of k″ part form H-bonds with residues of the strands in the other subsheet (strand a). We observed that k″ part always forms H-bonds only with the edge strand (strand a), as shown in Fig. 7.3b. This unique conformation of k strand allows for folding of the β-sheet into barrel structure. This long bent β-strand will be referred to as the "linking" strand. In the 1bia structure, strand #2 is the linking strand (Fig. 7.2).
(c) Some structures contain not one, but two neighboring H-bonded long antiparallel "linking strands." These two 90° bent strands cross over from one part of β-sheet to the other as shown in Fig. 7.2c. Residues in k′ and a″ will be found in one subsheet, while k″ and a′ will be found in the other one.
(d) Lastly, some barrel structures contain three linking strands. Typical relative orientation of the three strands in the main β-sheet is shown in Fig. 7.2d.

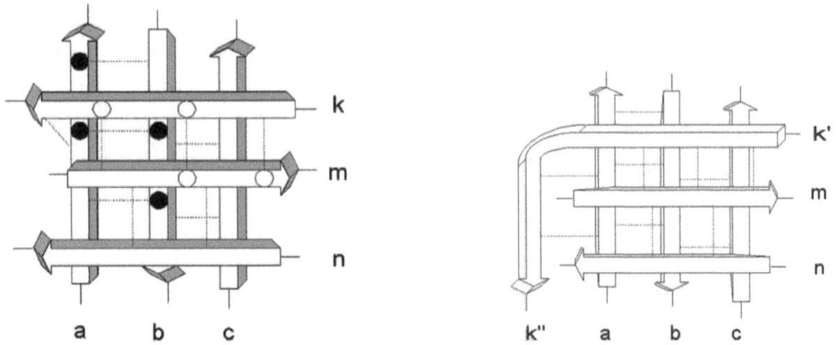

Fig. 7.3. The connection between the strands in two subsheets. (**a**) Two "short" edge strands in different subsheets form a hydrogen bond contact. The strands a, b, and c form one subsheet, while the strands k, m, n form another subsheet. The strands a and k are the edge strands in these two subsheets. (**b**) One "long" edge strand k forms H-bonds with the strands of both subsheets. The part at the beginning of the strand k′ is connected with the strand m of one subsheet and the part k″ is connected with the strand a of another subsheet. (**c**) Two "long" edge strands k and a form H-bonds with the strands of both subsheets. Residues of the k′ part of the strand k form H-bond contacts with the strand m, and a" parts of the strand a. Residues of the k″ part of the strand k form H-bond contacts with the strand a′. (**d**) Three "long edge strands k, m and a form H-bonds with the strands of both subsheets

7.4.7 Classification of Barrel Based on the Strands Arrangement

The arrangements of the barrel strands in two subsheets of the main β-sheet are shown in Table 7.2. It illustrates the arrangement of strands in the main β-sheet of one representative structure of each protein folds. For example, in the 1bia structure (fold # 40) the first subsheet contains mutually parallel strands $1, 2'$, and 5 (subsheet A), while the other subsheet is made of mutually parallel strands $2''$, 3, and 4 (subsheet B) (Fig. 7.1).

It was shown that all proteins of a given fold share the same arrangement of strands in the A and B subsheets. The arrangement of the strands in the A and B subsheets in a given fold is different from that of other fold. For example, the proteins in folds ## 48 and 49 have a similar arrangement of the strands in the β-sheet: 1-2-3-4-5-6; however the "strand compositions" of "β-subsheets" A and B are different (see Table 7.2).

It is important to mention here that in our research the secondary structure definition and the arrangement of the strands in the structures are based mostly on the analysis of the H-bond contacts between the main chain atoms. It follows that the information about the matrix of H-bond contacts between residues of a barrel structure is sufficient to formally assign a given barrel protein to its proper SCOP fold.

7.4.8 Characterizing the Place of Distortion of Barrel Structures

The place of distortion of the beta sheet is where the two orthogonal subsheets meet. It is largely defines the characteristic shape of the barrel proteins. It is reasonable to assume that the essential structural and sequence invariant features of barrels will be found at or near the place of distortion. We focus our attention, therefore, on the two "edge strands," a (or a′) and k (or k′) that bind the subsheets and their immediate strand neighbors (Fig. 7.3a–c). In one subsheet, strand a (or a') is H-bonded with strand b, while in the other subsheet strand k (or k′) forms H-bonds with strand m. Table 7.2 illustrates strand arrangements in various barrel folds; the four strands that make up the place of distortion are referenced by enlarged numerals in bold font.

The definition of the place of distortion as a part of the barrel structure containing two edge strands and their two neighbors requires a qualification. When we analyze the four strands of the place of distortion we omit from consideration the "double prime" portion of the barrel strand(s). It allows us to make our consideration general for all barrel structures. The strands that do not bend and the prime part of the link strand are the conserved elements of the barrel structures, while double prime portions of the link strands were found in the part of the barrel proteins.

7.4.9 The Rule of the Arrangement of the "Edge Strands" in the Barrel Structures

In 15 protein folds, the pair of strands in each subsheet that makes up the part of the distortion region are labeled by consecutive indices, such as the strands 2 and 3 in one subsheet and the strands 3 and 4 in the other in the structure 1bia (Fig. 7.2). Table 7.2 shows these constraints for the structures of the folds ## 40, 44, 45, 46, 48, 54, 65, 66, 52, 54, 55, 56, 59, 60, 69, 70, 71, 74, and 75 (see the enlarged in bold numbers of the strands).

In all other barrel folds one of the two subsheets contains a pair of sequentially numbered strands at the point of distortion (see subsheet B in Table 7.2). The exceptions of this rule are found only in structures of folds ## 55–57, where two subsheets are connected by three linking strands.

7.4.10 Arrangement of the Barrel and Sandwich Structures is Different

As shown above we found a strict constraint for the arrangement of the strands in the sandwich-like proteins. In almost all sandwich-like proteins there exists the definite rule of two interlocked pairs of strands, which are located in two β-subsheets. It is important to test this rule in the two subsheets of the barrel structure. The analysis of the arrangement of the strands in A and B β-sheets showed that interlocked arrangement does not exist in the barrel protein structures (the single exception was found in the proteins described in the protein fold # 57, "Domain of alpha and beta subunits of F1 ATP synthase-like" in SCOP database). It follows that the rule of two interlocked pairs of strands can serve to distinguish of the sandwich and barrel structures.

7.4.11 Invariant Substructure at the Place of Distortion: A Hydrophobic Tetrahedral

Analysis of residue content of the place of distortion revealed four conserved positions in each subsheet. The residues at each such position across the wide spectrum of barrel proteins all share certain sequence and structural properties. The four conserved positions of each subsheet can be said to represent the invariant substructure of their respective subsheet. A specific chemical characteristic of these positions is that at least three of the four conserved positions in each subsheet are occupied exclusively by hydrophobic residues.

If we take the four conserved positions in either subsheet to be the vertices of a closed geometrical figure then each subsheet will be seen to contain an imaginary tetrahedral. Interestingly, the invariant substructures of the two subsheets are essentially identical. The characteristic appearance of the invariant substructure is a consequence of the fact that the pairs of residues in the two strands that form the two opposite "faces" of the tetrahedral are located one reside away from each other. Figure 7.4 illustrates the invariant tetrahedral of each subsheet. Note that residues at the positions s in one strand and t in the other always share a hydrogen bond; the two upstream positions $s+2$ and $t+2$ complete the figure. A similar situation is obtained in the other subsheet, where the four tetrahedral positions are p and $p+2$, r and $r+2$, with an H-bond between residues at the p and r positions. Since at least three of the four positions of each tetrahedral are always hydrophobic, the invariant structures of the subsheets were termed "hydrophobic tetrahedras."

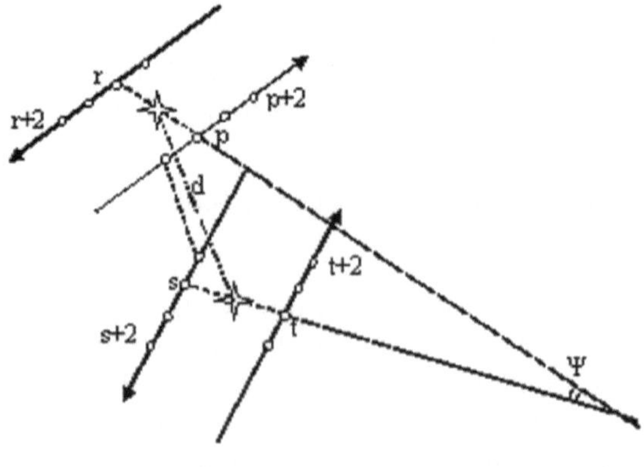

Fig. 7.4. Two tetrahedrals form "tetralock" in the barrel structures

7.4.12 The Two Hydrophobic Tetrahedrals Present the Structural Invariant of Barrel Proteins

The relative positions of the hydrophobic tetrahedral in the two subsheets can be described by two parameters: angle (ϕ) between the hydrogen bonds that connect the residues at positions S and T in one tetrahedral and positions P and R in the other, and distance between the "centers of gravity" of the two tetrahedrals (Fig. 7.4). The pair of tetrahedral makes up the invariant substructure of the barrel proteins. Each barrel protein contains at least one pair of tetrahedral, but some contain several such pairs with approximately same distances between centers of gravity and angles.

7.5 Conclusion

The investigation carried out thus far allowed us to find the invariant sequence and structural invariants in the vast diversity of beta structures. Analysis of the supersecondary structures revealed the constraints in the arrangements of the strands. It supports the conclusion that proteins which grouped together on the basis of common architecture like sandwich-like or barrel-like proteins have commonality on the level of supersecondary structure.

In both groups of proteins – sandwich and barrel structures the arrangement of strands gives two invariants substructure – hydrophobic tetrahedrals. In fact, the tetrahedral, that make up of residues of neighboring strands in the beta sheet can be considered as the structural unit of the beta structures.

In the sandwich-like structure two tetrahedrals form interlock. The residues of the interlock lie at the center of the interface between the beta sheets and form the common geometrical core of sandwich proteins.

In the barrel structure two tetrahedrals form another geometrical figure – tetrahedral. Residues of the tetralock lie at the edge of two subsheets. We can suggest that these residues are responsible for the distortion of the β-sheet in the barrel-like structure, which results in the formation of two subsheets and leads to the closed structure. The tetralock can be considered as the common geometrical core of the barrel proteins.

Analysis of a broad groups of proteins, such as sets of superfamilies, yields a set of sequence determinants of a group of nonhomologous proteins. These sequence determinants form the basis of computer algorithm for classification of novel proteins.

A direct corollary of our approach is that complexity of protein sequence search algorithms and 3D structure predictions can be dramatically reduced: instead of carrying out searches with whole protein sequences, we may now carry out searches with predefined sets of several key residues. This is analogous to searching for a suspect by his fingerprints, rather than by a long list of nonunique descriptors. Our data on sandwich-like proteins shows that the proposed search algorithm compares favorably with the powerful and widely used techniques based on hidden Markov Models.

Another advantage of carrying out a structure-based analysis is that it often allows one not only to predict the affiliation of a particular protein and outline its secondary and 3D structure, but also to make "educated guesses" about functional role of various portions of the sequence. It is evident that an ability to pinpoint parts of protein sequence that is likely to take part in protein binding, for example, can prove invaluable for planning mutagenesis experiments or rational drug design.

References

1. National center for biotechnology information. http://www.ncbi.nlm.nih.gov/
2. H.M. Berman, J.Z. Westbrook, Z. Feng, G. Gilliland, T.N. Bhat, H. Weissig, I.N. Shindyalov, P.E. Bourne, Nucleic Acids Res. **28**, 235–242 (2000). http://www.rcsb.org/pdb/
3. A.G. Murzin, S.E. Brenner, T. Hubbard, C. Chothia, J. Mol. Biol. **247**, 536–540 (1995). http://scop.mrc-lmb.cam.ac.uk/scop/
4. C.B. Anfinsen, Science **181**, 223–230 (1973)
5. E.E. Lattman, G.D. Rose, Proc. Natl Acad. Sci. USA **90**, 439–441 (1993)
6. T.C. Wood, W.R. Pearson, J. Mol. Biol. **291**, 977–995 (1999)
7. A.A. Schaffer, L. Aravind, T.L. Madden, S. Shavirin, J.L. Spouge, Y.I. Wolf, E.V. Koonin, S.F. Altschul, Nucleic Acids Res. **29** (14)2994–3005 (2001). http://www.ncbi.nlm.nih.gov/BLAST/
8. W.R. Pearson, D.J. Lipman, Proc. Natl Acad. Sci. USA **85**, 2444–2448 (1988)
9. J. Park, K. Karplus, C. Barret, R. Hughey, D. Haussler, T. Hubbard, C. Chothia, J. Mol. Biol. **284**, 1201–1210 (1998)
10. C. Chothia, A.M. Lesk, EMBO J. **5**, 823–826 (1986)
11. C. Chothia, A.M. Lesk, I.M. Gelfand, A.E. Kister, in, *Why Structure Changes More Slowly Than Sequence in Protein Evolution, Simplicity and Complexity in Proteins and Nucleic acids* ed. by H. Frauenfelder, J. Deisenhofer, P.G. Wolynes. (Dantem University Press 1999), pp. 281–295
12. C. Chothia, I.M. Gelfand, A.E. Kister, J. Mol. Biol. **278**, 457–479 (1995)
13. C. Chothia, T. Hubbard, S. Brenner, H. Barns, A. Murzin, Biomol. Struct. **26**, 597–627 (1997)
14. C.A. Orengo, A.D. Michie, S. Jones, D.T. Jones, M.B. Swindells, J.M. Thornton, *Structure* **5**(8), 1093–1108 (1997)
15. J.S. Richardson, Adv. Protein Chem. **34**, 167–339 (1981)
16. J.S. Richardson, D. Richardson, Natl Acad. Sci. USA **99**, 2754–2759 (2002)
17. A.D. McLachlan, J. Mol. Biol. **128**, 49–79 (1979)
18. I. Ruczinski, C. Kooperberg, R. Bonneau, D. Baker, Proteins **48**, 85–97 (2002)
19. A.G. Murzin, A.M. Lesk, C. Chothia, J. Mol. Biol., **236**, 1369–1381 (1994a)
20. A.G. Murzin, A.M. Lesk, C. Chothia, J. Mol. Biol., **236**, 1382–1400 (1994b)
21. N. Nagano, C.A. Orengo, J.M. Thornton, J. Mol. Biol. **321**, 741–765 (2002)
22. R.E. Steward, J.M. Thornton, Proteins **48**, 178–191 (2002)
23. F.A. Syud, H.E. Stanger, H.S. Mortell, J.F. Espinosa, J.D. Fisk, C.G. Fry, S.H. Gellman, J. Mol. Biol., **326**, 553–568 (2003)

24. C. Zhang, S.-H. Kim, Proteins: Struct. Funct. Gen. **40**, 409–419 (2000a)
25. C. Chang, S.-H. Kim, J. Mol. Biol., **299**, 1075–1089 (2000b)
26. W.-M. Liu, Protein Eng., **10**, 1373–1377 (1997)
27. W.-M. Liu, J. Mol. Biol., **275**, 541–545 (1998)
28. G.M. Salem, E.G. Hutchinson, C.A. Orengo, J.M. Thornton, J. Mol. Biol. **287**, 969–981 (1999)
29. C. Chothia, A.V. Finkelstein, Annu Rev. Biochem. **59,** 1007–1039 (1990)
30. D.N. Woolfson, P.A. Evans, E.G. Hutchinson, J.M. Thornton, Protein Eng. **6**, 461–470 (1993)
31. P.J. Kraulis, J. Appl. Crystallogr. **24**, 946–950 (1991)
32. A.E. Kister, M.A. Roytberg, C. Chothia, Y.M. Vasiliev, I.M. Gelfand, Protein Sci. **10**, 1801–1810 (2001)
33. A. Kister, A. Finkelstein, I. Gelfand, Proc. Natl Acad. Sci. USA **99**, 14137–14141 (2002)
34. J. Gough, C. Chothia, Nucleic Acids Res. **30**(1), 268–272 (2002). http://supfam.mrc-lmb.cam.ac.uk/SUPERFAMILY/index.html

8

The Structure of Collagen

N. Rivier and J.-F. Sadoc

Summary. We study the assembly of collagen molecules of the so-called fibrils, long, periodic bundle of finite collagen molecules. The appearance of three-dimensional periodic structures leads to very interesting geometrical questions similar to the problems of classification textures and defects in liquid crystals (smectics and discotics), lattices of defects in superconductors, defects in liquid membranes, dense packing of spheres, and so on.

8.1 Collagen: Chain, Molecule, Fibril

Collagen is the principal constituent of extracellular, connective tissue. It is made of fibrils, which are close-packed bundles of long molecules. Each molecule consists of three intertwined polypeptide chains, forming a right-handed helix (Fig. 8.1b, c). The chain is a nearly periodic sequence of amino acids (a.a) $\ldots. - [Gly - X - Y] - \ldots$, where the a.a X and Y are predominantly Pro or $HPro$. Collagen is a protein, but with a periodic (period 3 in Gly) primary structure, and a helix that is altogether its sole secondary structure and its ternary structure (Fig. 8.1a). The helix has 2. 73 a.a per turn, in contrast with the 3.6 of the pervasive alpha helix. Thus, collagen is a protein that is only a material, and this chapter is not an attempt to inject some mathematics or physics into biology, but a recognition that some biological constituents are simply material science.

The collagen molecules assemble into fibrils. Longitudinally, the molecules are separated by gaps, and the fibril is a periodic alternance of overlap and gap regions, as indicated in Fig. 8.1d. The transverse structure is, in both regions, a topological Archimedean square–triangle lattice $3^2.4.3.4$, also known as the main skeleton of the Frank–Kasper sigma phase, and observed by Okuyama et al. [1] in $[Gly - Pro - Pro]_{10}$. The three-dimensional structure is a rotating stack of successive overlap–gap–overlap..., separated by boundaries of twist dislocations. The stack is periodic, as is observed.

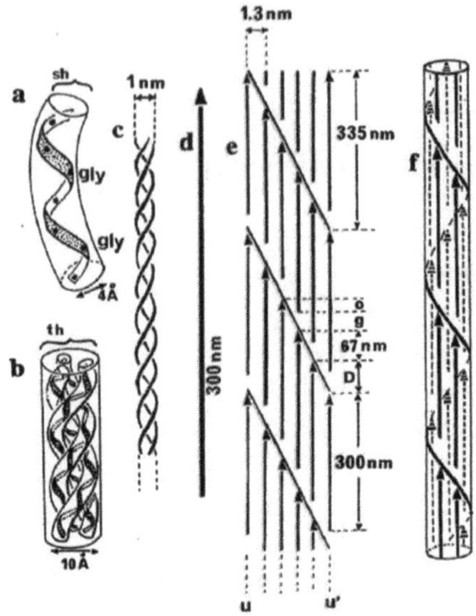

Fig. 8.1. Structure of collagen at various scales. (**a**) Single chain $Gly - X - Y$ forming a left-handed helix, with a curved axis. *Dots* indicate the positions of successive amino acids. Their side groups are pointing outwards; every third one is a glycine (gly) located in the concave side of the axis. (**b**) A collagen molecule is a right-handed triple helix of three intertwined polypeptide chains, each one being left-handed as in (**a**). The core of the triple helix consists of the side groups of the glycines, a tightly packed helix of H atoms. (**c**) A simpler representation of the right-handed triple helix, represented more simply. Its width is 1 nm, its length 300 nm. (**d**) The collagen molecule is often represented by an arrow of length 300 nm. (**e**) The molecules are stacked on top of each other, with a gap of 35 nm, and regularly spaced on a lattice (transverse scale dilated relative to longitudinal scale). The longitudinal coordinates of molecules u and u' are the same. One distinguishes gap and overlap levels, g and o, respectively, so that to each five molecules in o correspond only four in g (the extremities of the molecule lie at gap–overlap interfaces). There is a regular stagger of molecules by a length $D = g + o = 67$ nm. (**f**) A crude representation of the lattice as a cylinder (by identifying u and u'). This is only schematic as the collagen fibril is a lattice, showing the 67-nm stagger, but with a unit cell that accommodates gap and overlap levels. Notice the right-handed chirality of the *Gly* core (see also Fig. 8.2 and Fig. 8.4b) and of the triple-helix collagen molecule, opposite to the left-handed chirality of the single collagen chain (polyproline II: PPII; see also Fig. 8.3). In Fig. 8.6 of [4], the PPII helix is wrongly drawn as right handed. We regret this oversight: The figure, obtained from some public domain file, was added at the proof stage

The essential physical and geometrical features to be included in the structure of the collagen fibril are:

(a) Close packing of amino acids in a bundle of periodic, polypeptide chains
(b) Flexibility of the fibril, compatible with close packing

(c) Fibrils of arbitrary length
(d) Intertwining of the chains within the molecule at a constant rate; this rate of intertwining does not decrease as the molecule gets longer

Points (c) and (d) suggest a pattern of molecules along the fibril, and of amino acids within the collagen molecule, that is either periodic or inflationary (quasicrystalline).

Diffraction, especially that of crystalline $[Gly - Pro - Pro]_{10}$, indicates that the pattern is periodic [1–3], but a chain of amino acids packs naturally as a Boedijk–Coxeter helix, that contains several approximants of $1 + \sqrt{3}$ [4] (see Sect. 8.2). These observations are not incompatible. The (longitudinal and transverse) structure of collagen is ultimately periodic, but its unit cell exhibits many successive approximants.

8.2 The Boerdijk–Coxeter Helix and its Approximants

Helices and densely packed spherical objects are two closely related geometrical problems. The simplest means of packing tightly a chain of connected spheres (representing amino acids) of arbitrary length is the Boerdijk–Coxeter (B–C) helix, represented in Fig. 8.2. It is a stacking along one direction of regular tetrahedra, the elementary unit of four close-packed spheres. Figure 8.1c represents the helix as a two-dimensional graph on triangular lattice covering

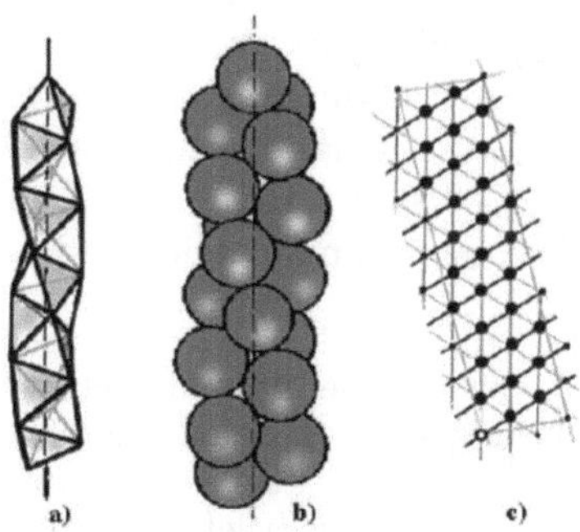

Fig. 8.2. Boerdijk–Coxeter helix (left-handed) obtained from a necklace of tetrahedra (**a**) or as a packing of spheres (**b**). In (**c**), a right-handed B–C helix is represented on a flat strip tiled with equilateral triangles, that constitutes a cylinder with the vertical grey lines identified. This is a 30/11 helix (30 vertices for 11 turns, a convergent of $1 + \sqrt{3}$). Note that the axis of the helix (*vertical grey line*) is not exactly perpendicular to its base (*horizontal grey line*). When the horizontal grey lines are also identified, one obtains one of the four tori that make up the Hopf fibration of polytope $\{3, 3, 5\}$, a discrete representation of the hypersphere S^3. There are three fibres–the steepest lines of ten neighbouring vertices, great circles in S^3 winding 1:1 around the torus

a cylinder. The cylinder is cut and flattened. This elegant geometric construction cannot be considered as a material unit for two reasons:

(a) The regular tetrahedron is not a three-dimensional space filler. Indeed, its dihedral angle is $2\pi/(5.1)$, and that is geometrically frustrated because the .1 leaves empty space, and the fivefold symmetry is not crystallographic. Indeed, the perfect, close packing of spheres exists in the positively curved space S^3; it is polytope $\{3, 3, 5\}$, containing 120 spheres and 600 tetrahedra.

(b) As a one-dimensional structure, it is not periodic, not even quasi-periodic, that is extensible from a small finite nucleus by substitution rules or by cut-and-projection through a sequence of approximants. The number of edges or of spheres per turn of the helix[1] is $2\pi/\cos^{-1}(-2/3) = [2,1,2,1,2,1,1,2,1,1,7,6,\ldots] \approx 2.7312$, that is neither a rational number, describing a periodic structure, nor a quadratic irrational, necessary condition for context-free inflation–deflation symmetry.[2]

It is possible to resolve the difficulty (b). One can construct a quasicrystalline Coxeter helix with a number of spheres per turn equal to $1 + \sqrt{3} = \underline{[2,1]} = 2.73205$, with exactly the same rational convergents through the first 112 amino acids and 41 turns. Notably, the periodic helices $30/11 = [2,1,2,1,2] = 2.7272\ldots$ of Fig. 8.2c), and $41/15 = [2,1,2,1,2,1] = 2.733\ldots$ are rational convergents of the B–C and quasicrystalline Coxeter helices. Neither have axis perpendicular to the base of the cylinder; as befits successive rational convergents, it lies on either side of the axis of the quasicrystalline Coxeter helix, which has an irrational slope in the underlying triangular lattice. There exists a third helix, with its axis exactly perpendicular to the base of the cylinder. It has 14 amino acids for 5 turns exactly, i.e. $42/15 = 14/5 = [2,1,4] = 2.8$ that has the same convergents (principal and intermediate) through the first 11 amino acids and 4 turns.[3] This is the basic helix of collagen.

---

[1] Three types of helical chains of spheres in contact, with different pitch and chirality, can be distinguished in the B–C helix. We refer here to the flattest, right-handed helix, called Coxeter chain. Each one of the three polypeptide chains constituting the collagen molecule, is a left-handed helix of intermediate pitch (Fig. 8.3), also constructible on the B–C helix. The *Gly* sit on the steepest, right-handed helix, that is also a fibre in the Hopf fibration of polytope $\{3,3,5\}$.

[2] A number is represented by its continuous fraction, as $q_0 + \frac{1}{q_1 + \frac{1}{q_2 + \ldots}} = [q_0, q_1, q_2, \ldots]$. A quadratic irrational has a periodic continuous fraction expansion, with the period underlined, e.g. $\sqrt{3} = 1 + \frac{1}{1 + \frac{1}{2 + \frac{1}{1 + \ldots}}} = [1, \underline{1,2}]$. Rational approximants are given by truncation of the sequence [5].

[3] In the collagen chain (Fig. 8.3), the axis goes through $Gly(0)$, $X(7)$, $Y(14)$ and $Gly(21)$. The (left-handed) helix with an axis perpendicular to the base is 21/6, or 7/2 for identical vertices. It has been identified by Okuyama et al. [1].

The geometrical frustration (a) is resolved in (positively) curved space [6]. Hopf's fibration of S^3 extends to its discrete scaffolding, polytope $\{3,3,5\}$ [7]. There are 12 fibres of ten vertices each. They can be divided into four co-axial tori of three intertwined fibres. All fibres are identical, by definition, and also equidistant (parallel in curved space). One torus flattened in Euclidean space is represented in Fig. 8.2c. It is covered by helix 30/11. One notices the three fibres, the steepest helices winding 1:1 around the torus. Note that edges of the triangular network are only equal in curved space. In Euclidean space, there is a slight distortion of the distance between different fibres. The base space is the icosahedron 3^5 (Fig. 8.4). Each vertex is the representative (a projective map) of one fibre.[4]

Incidentally, the torus covered by helix 42/15 can also be represented by a cylinder, with its axis perpendicular to the base. The three fibres winding 1:1 around the torus are longer: They have 14 vertices each. These fibres are identical to the four fibres in the torus covered by a disclinated Coxeter helix $1 + 41/15$, based on a square rather than on a triangle [4]. This enables us to extend the Hopf fibration of polytope $\{3,3,5\}$ to a decurved (flattened) polytope of $24 \times 14 = 336$ vertices, with fibres of 14 vertices instead of ten. And then, without any further lengthening of the fibres, to the Euclidean honeycomb with the sigma phase honeycomb $3^2.4.3.4$ as base space.

8.3 The Collagen Molecule

The collagen molecule, with its three, intertwined polypeptide chains, can be represented in curved space on the Hopf fibration of polytope $\{3,3,5\}$. The base space is an icosahedron.

A single collagen chain (polyproline PPII helix structure) forms a left-handed helix $[Gly - X - Y]_5$, (15/4), drawn in Fig. 8.3 on a torus of the fibration, i.e. on the right-handed B–C helix 30/11, with twice the vertical spacing between vertices. Half of the vertices on the torus, constituting a second left-handed helix parallel to the first, are empty. The fibres (of the original Hopf fibration, with all vertices identified) are the three, steepest right-handed helices. One contains five Gly separated by five empty vertices. The other two, five amino acids X or Y, respectively, separated by five empty vertices. The torus is mapped on the base space as a triangle (shaded in Fig. 8.4b).

The collagen molecule is represented as a hexagon on base space, with a triangular core of Gly inside (Fig. 8.4b). Within the triangular core, the two hydrogen side groups of Gly form a close-packed Coxeter helix 30/11 (with the 30 H attached to the 3×5 Gly from the three chains). The projection of the horizontal, interchain, hydrogen bonds are represented as double lines in Fig. 8.4b.

[4] We have to distinguish between base space of a non-trivial fibration and base of a cylinder (a circle).

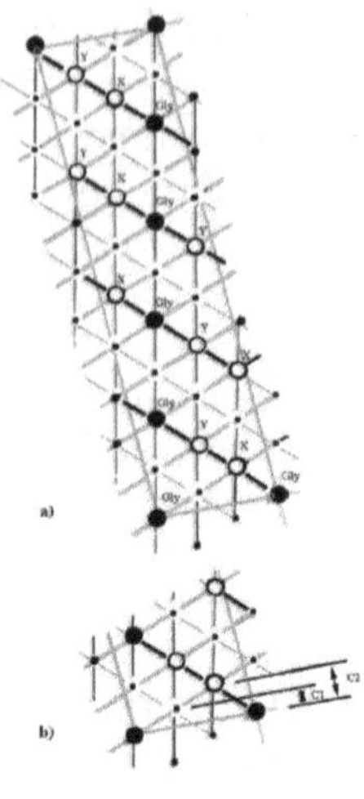

Fig. 8.3. The single collagen chain PPII ($Gly - X - Y$) is a left-handed helix, that can be drawn on a right-handed B–C helix with half the vertical spacing. The helix is represented on a flat, rectangular strip by identification of the two long sides of the rectangle. In the Hopf fibration of polytope $\{3, 3, 5\}$, five amino acids Gly, (or X, or Y, resp.) lie on a fibre that is a great circle with ten vertices. Note that the axis $Gly - X$ of helix 7/2 is orthogonal to the base, so that a stack of three segments 7/2 gives one period of the collagen chain in Euclidean space, the left-handed helix 21/6, with a fibre of seven Gly winding around once. The winding number of the underlying right-handed B–C helix is 41/15, the next convergent, after 30/11 (Fig. 8.2c), of $1 + \sqrt{3}$. There are 14 vertices in the fibre. The collagen chain is periodic and the embedding space is completely decurved

8.4 Decurving

The structure of collagen in Euclidean space is obtained by decurving space, i.e. by iteratively increasing the radius of the polytope, in a way that keeps constant the rate of intertwining of the chains, and the connectivity and symmetry of the base space. Decurving increases the radius and area of the base space and lengthens the fibres. In a fibration, the base must remain an Archimedean polyhedron or honeycomb, a tiling of triangles and squares (decurved triangles) with vertex connectivity $z = 5$.[5] This suggests a decurving of the base

[5] If the individual helices are represented by triangles in base space, as in Fig. 8.4b, decurving must not involve any physical distance. The triangle that changed into a square is neither the core of three Gly, nor the triangle $Gly - X - Y$ representing one polypeptide chain, nor even the triangle $Gly - (Gly - X)$, because it involves the horizontal hydrogen bond between the Gly of one chain and the X of another. It can only be the triangle $(Y - Gly) = X$, with the distance between the X of one chain and the Y of the other stretched to become the diagonal of the square. The snub cube contains then the projection of two complete molecules. If the triangle representing the polypeptide chain was changed into a square, one would have obtained an alpha helix [4].

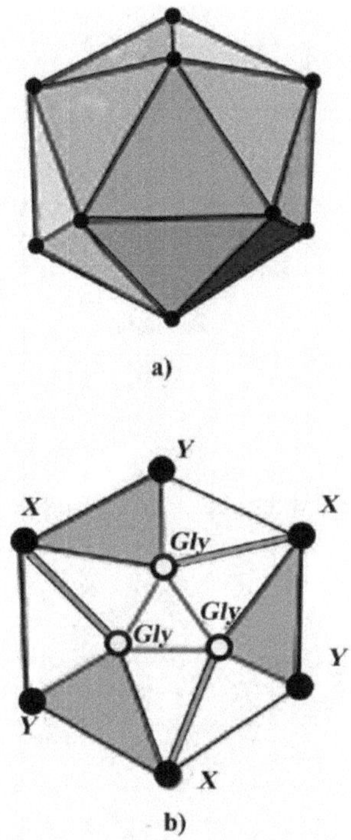

Fig. 8.4. The base space of the Hopf fibration of polytope $\{3, 3, 5\}$ is an icosahedron (**a**). Each vertex of the icosahedron is only the representative of one fibre. Physically, the fibre is either a collagen molecule (a triple helix), or (**b**) the representative of the amino acids *Gly* (or X, or Y, resp.) of a single PPII chain. Then, *a shaded triangular face* of the icosahedron represents a PPII chain, and the triple helix collagen molecule is represented by the decorated hexagon in (**b**). Hydrogen bonds (*double lines*) are horizontal bridges between the *Gly* of one helix and the X of another. Notice the right-handed chirality of the *Gly* core and of the triple collagen molecule, opposed to the left-handed chirality of the PPII helices

in two stages, icosahedron 3^5 to snub cube $3^4.4$, to sigma phase honeycomb $3^2.4.3.4$.

Alternatively, one can represent an entire molecule as a vertex in the base space. This alternative representation is used hereafter (Fig. 8.5–8.9).

The first step decurves the base space of $\{3, 3, 5\}$, an icosahedron 3^5 into a snub cube $3^4.4$. The snub cube can be decurved one step further, by replacing a second triangle by a square. One obtains the square–triangle, Archimedean, $z = 5$ honeycomb $3^2.4.3.4$ (major skeleton of the sigma phase of intermetallic compounds, a Frank–Kasper phase).[6] Decurving is complete, because the base space is now flat and infinite, and the fibres are periodic B–C helices 42/15.

6 The alternative honeycomb $3^3.4^2$, in which the two squares are neighbours, forfeits the isotropy of the original polytope $\{3, 3, 5\}$.

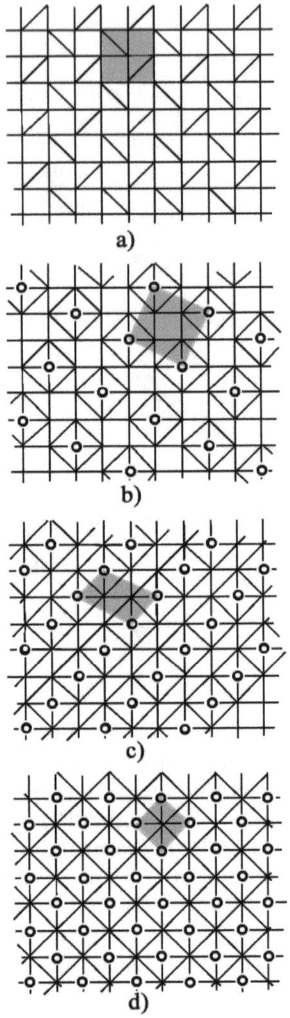

Fig. 8.5. Bouligand's overlap–gap transformation between two Archimedean lattices. Unit cells in grey. (**a**) A square lattice decorated as a topological square–triangle lattice $3^2.4.3.4$ (overlap, $z = 5$). (**b**) By removing one vertex out of five from the square lattice, one obtains a topological square–triangle lattice $3^2.4.3.4$ (gap, $z = 5$). The square unit cell has area $\sqrt{5} \times \sqrt{5} = 5$. The vertices removed from the original lattice are noted as **o**, and are replaced by a diamond square. (**c**) The other Archimedean, $z = 5$ alternative is $3^3.4^2$, but it is much less isotropic. The unit cell is not square, with an area $\sqrt{2} \times (\sqrt{2} + 1/\sqrt{2}) = 3$, and one vertex out of three has been removed. (**d**) The only Archimedean alternative is between square lattices ($z = 4$). The square unit cell has area $\sqrt{2} \times \sqrt{2} = 2$ (one vertex out of two has been removed)

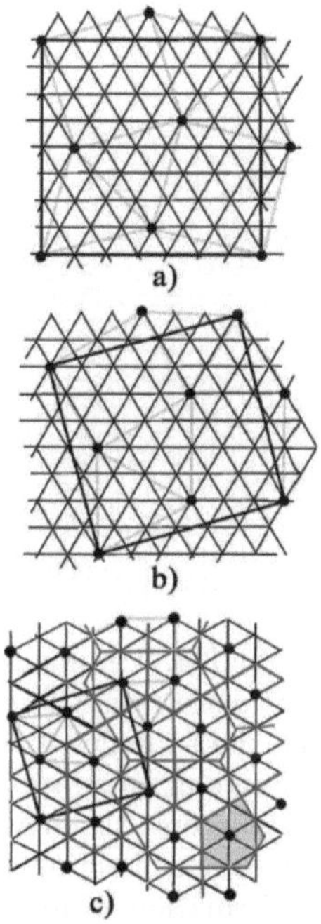

Fig. 8.6. (**a**) The unit cell of the square–triangle $3^2.4.3.4$ gap structure, drawn on a triangular lattice. All edges have equal length, but the "square" is in fact a rhombus. A vertex is the representative of a collagen molecule. (**b**) The unit cell of the square–triangle $3^2.4.3.4$ overlap structure, drawn on a triangular lattice. The "square" is slightly rectangular, and there are two types of equilateral triangles, of sizes in the ratio $2\sqrt{3}/3 = 2/\sqrt{3}$. (**c**) A smaller $3^2.4.3.4$ overlap structure, with the square of the gap structure as unit cell, drawn on a triangular lattice, rotated by $\pi/2$. It is similar to the original overlap structure (**b**). The factor of similarity is $1/\sqrt{3}$. Also drawn (grey) is the associated Voronoi tiling

Moreover, the axis of the helix is perpendicular to its basis, and the projection is orthogonal. The structure remains a fibration at all stages.

The identical fibres are helices winding 1:1 around the tori represented by the triangles and squares of the base space. They have ten vertices in the original $\{3, 3, 5\}$ polytope (with the tori covered by the Coxeter helix 30/11, Fig. 8.2c). At the next two stages, the fibres are longer (14 vertices) but they still wind 1:1 around the longer tori: The torus represented as a triangle in base space is covered by the Coxeter helix 42/15. Torus and helix are periodic in Euclidean space (the axis of the strip making up the torus with opposite sides identified, is perpendicular to the base).

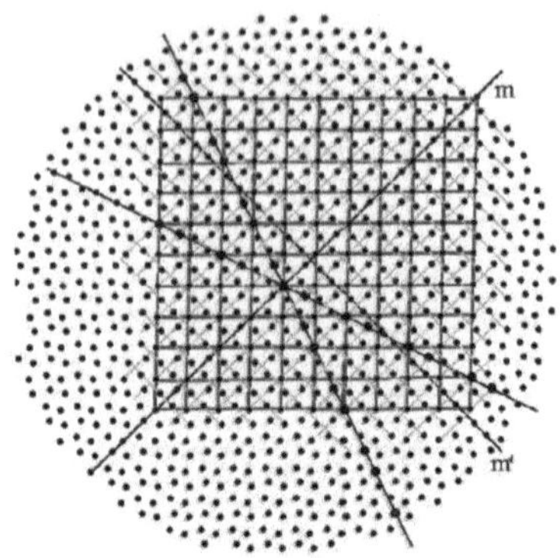

Fig. 8.7. The overlap structure $3^2.4.3.4$. The main grid is that of the unit cell (a square with a slight rhombohedral distortion). The secondary grid is made of the mirror planes m and m', bisecting the triangles. It has a rectangular unit cell with half the area. There are two other strong alignments, on the directions $(2, -1)$ and $(-1, 2)$ of the main grid, that include one diagonal of one of the two squares and the intersection between two mirror lines bisecting the larger triangles. The aligned points are equidistant. The transformation from one alignment to the other is a rotation of $\simeq 27° = (3/2)(\pi/10)$. The overlap structure has many symmetries (unlike the gap), responsible for the periodic, twist grain boundary stacking of sections of intertwined collagen molecules that constitute the fibril of collagen

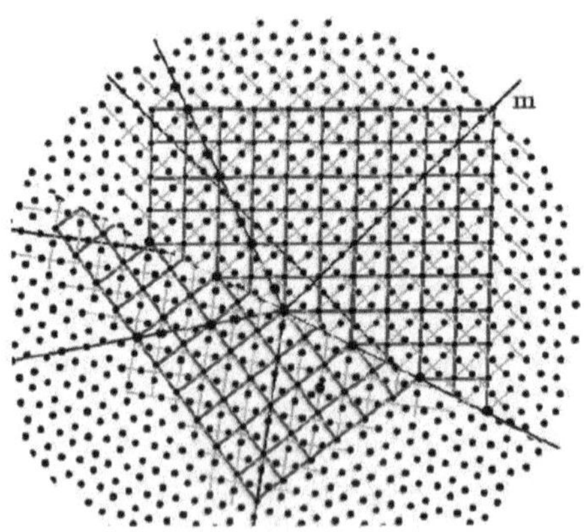

Fig. 8.8. The twist grain boundary between two overlap structures at successive heights, rotated by $\simeq 27°$. Note the periodic coincidence site lattice between equidistant, aligned points

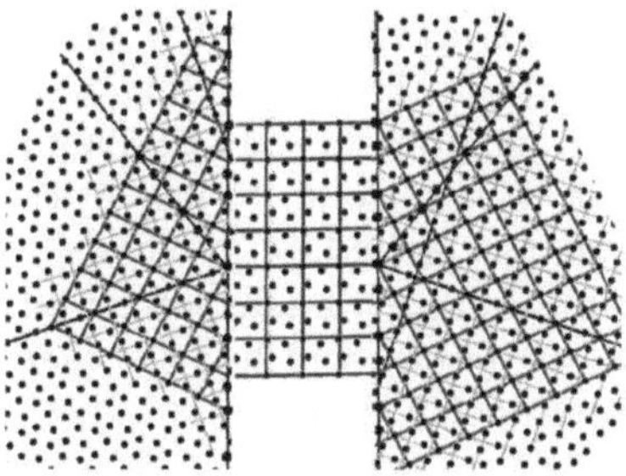

Fig. 8.9. Two successive twist grain boundaries overlap–gap–overlap. This illustrates the 5/4 ratio between the areas of the overlap and gap unit cells. The structure of a collagen fibril is thus a periodic stacking of twist grain boundaries $[overlap - gap]_{10}$

8.5 Transverse Structures (gap, overlap) on Two Orthogonal Triangular Lattices

Since the collagen helix is represented on a cylinder covered by a triangular lattice (Fig. 8.2c and 8.3), and since it is inflated by $1 + \sqrt{3}$, the transverse structure of the collagen fibril should be based on an underlying triangular lattice.

The transverse structure is, topologically, a square–triangle pattern $3^2.4.3.4$, with vertex coordination $z = 5$. This pattern accommodates the flexibility of tightly packed individual molecules through gaps in their stacking (Fig. 8.1d). Accordingly, the transverse structure at the "gap" level is topologically the same as at the "overlap" level, but with density reduced by 4/5, and overall rotation (see Fig. 8.5b).[7] The transition overlap–gap–overlap is a twist grain boundary (TGB). In general, the one-dimensional stacking of TGB is periodic [9], but, in collagen, this periodicity, and the square–triangle transverse structure, are necessary consequence of close packing of amino acids, that is of steric repulsion.

The mechanism of the transition from overlap to gap is illustrated in Fig. 8.5 [10], showing how an interruption in one out of every five molecules leaves the square–triangle lattice invariant. The topological rotation is by $\tan^{-1}(1/2)$. Figure 8.5 also shows that the square–triangle ($z = 5$) and square ($z = 4$) patterns are the only topological lattices invariant through regularly spaced gaps. Each vertex has a label $1, 2, 3, 4, 5$ (mod.5) that indicates the gap level at which its representative molecule is interrupted. The numbers increase by 3 (mod.5) horizontally, and by 1 vertically, in the positive sense (by 2 and by 4 in the negative sense). The interrupted molecule in the gap region has four neighbours with labels all different. Every vertex in the gap region has five neighbours, with labels different from its own and from that of the gap.

[7] The fact that out of the five molecules in the overlap, only four extend into the gap has been established by Hodge and Petruska [8]. See [3, 10].

Let us now obtain the metric of the two structures. The inflation multiplier $1+\sqrt{3}$ must be present both in the longitudinal and in the transverse structures, which is expected to exhibit several rational approximants of $\sqrt{3}$. This suggests constructing the transverse structure on two orthogonal triangular lattices. The superposition pattern exhibits "wheels" at rational approximants of $1+\sqrt{3}$, located on (two) square–triangle pattern(s). The wheels are centred on two copies of the gap structure, which is the main coincidence pattern of the two perpendicular, triangular lattices.

8.6 The Gap Structure

The gap structure is shown in Fig. 8.6a. All edges are equal. But the "square" has a slight rhombic distortion, and the unit cell (4 vertices, 4 triangles, 2 squares, with an area of 112 elementary triangles) is slightly rectangular. The distortion is equal to $4\ \sqrt{3/7} = 1.010$. (see Table 8.1). The gap structure is metrically regular, but it has no mirror symmetry, and no alignments of points.

8.7 The Overlap Structure

The overlap structure is shown in Fig. 8.6b. It is built on rectangular "squares" with orthogonal sides on the 12 symmetry directions of the underlying triangular lattice, and with 5/4 the vertex density of the gap structure. The arithmetic construction of the gap and overlap structures is detailed in the next section.

Table 8.1. The table below lists the geometric manifestations of the various approximants C_j/A_j to $\sqrt{3}$ in the overlap and gap structures. A rectangular structure is denoted as [...]; it has one approximant . Without bracket, it refers to an equilateral rhombus, with two approximants (a) and (b). The unit cell of the small overlap structure is the square of the gap structure. The first column gives the value of j. The last column gives the (orthorhombic or triclinic) distortion, $\sup(\sqrt{3}/\text{app.}, \text{app.}/\sqrt{3})$ For reference:
$j = 4\ \sqrt{3} \approx 19/11$ in B.C. helix.
$j = 5\ \sqrt{3} \approx 41/15$ in inflated B.C. helix.

j	$\sqrt{3}$	overlap	small lap	over-gap	distortion
1	2/1	[square]	[square] unit cell(a)		
		unit cell(a)	unit cell(a)	square(a)	1.155
2	5/3	[mirror unit cell]	[mirror unit cell]		1.04
3	7/4	unit cell(b)	unit cell(b)	square(b)	1.01
			[unit cell]		

The overlap structure is much more symmetrical than the gap. It has:

(a) Orthogonal mirror axes that coincide with the diagonals of the rhombi.
(b) Many points are aligned and regularly spaced, on the direction of the diagonal of the square, which is also direction (2, −1) in the unit cell grid.
(c) It has two unit cells, a rhombus-shaped unit cell (with four vertices and an area of 90 elementary triangles)[8], and a smaller, rectangular unit cell made of mirror planes (two vertices).
(d) It exists at two different scales (Fig. 8.6b, c) with the same convergents.

One can constitute a "square" cell in the overlap structure, of sides (2,−1) and (−1,−2), containing 20 points (or 5 unit cells, arranged as in Fig. 8.7). A square of 2×2 unit cells of the gap structure contains 16 points. The two cells fit almost exactly (side length: $\sqrt{196}$ and $\sqrt{192}$ (gap); $\sqrt{189}$ (overlap)). The rotation $\tan^{-1}(\sqrt{3}/6) + \tan^{-1}(\sqrt{3}/9) = \tan^{-1}(5\sqrt{3}/17) = 26.996° = 3\pi/20$ is the metric equivalent of the topological rotation of $\tan^{-1}(1/2)$ of Bouligand. The small orthorhombic distortions are different in the gap and in the overlap. This is why one can observe two superposed, different X-ray diffractions patterns, corresponding to gap and overlap distances. Experimentally, the distortion is 2% on average [1].

8.8 Transverse Structure; Coincidence Lattice of Two Orthogonal, Triangular Lattices; Approximants of $\sqrt{3}$

The inflation multiplier for the collagen molecule is $1 + \sqrt{3} = [2, \overline{1}]$. It is expected to dominate also the transverse structure of collagen, that should be based on the coincidence points of two perpendicular triangular lattices, with coordinates corresponding to rational convergents of $\sqrt{3}$. The superposed triangular lattices have a non-crystallographic, 12-fold rotation symmetry. The resulting lattice of near-coincidence points is, topologically, (two copies of) the Archimedean lattice $3^2.4.3.4$. It is the gap structure, with unit cell shown in Fig. 8.6a. It has nearly square symmetry, with a slight orthorhombic distortion of 1.01.[9]

A triangular lattice is the set of points $(1/2)a\boldsymbol{i} + (\sqrt{3}/2)b\boldsymbol{j}$, with a, b integers, both odd or even, where $\boldsymbol{i}$ and $\boldsymbol{j}$ are orthogonal unit vectors. The 12 axes of symmetry are given by $a = \pm\, b$ or by $a = 0$ or $b = 0$, and permutation of $\boldsymbol{i}$ and $\boldsymbol{j}$. If the origin is one coincidence point, another lies:

---

[8] The areas of the primitive cells match: overlap: $(5/4)90 = 112.5$, gap: 112 elementary triangles.

[9] Two triangular lattices, rotated by θ and superposed, give moiré patterns in general, except for $\theta = \pi/3$ (exact superposition) and $\theta = \pi/2$ (mod.$\pi/3$) (two Archimedean lattices $3^2.4.3.4$).

(1) On an axis of 12-fold symmetry. It is given by the lattice vectors $n\,\boldsymbol{i} \approx p\sqrt{3}\,\boldsymbol{i}$ in the two, orthogonal triangular lattices. Thus, $\sqrt{3} \approx n/p = C_j/A_j$, where $C_j = B_j A_j$ and B_j/A_j are successive convergents of $1+\sqrt{3}$, obtained by truncation of its continued fraction.[10] The A_j are given by recursion, $A_j = q_j A_{j-1} + A_{j-2}$, with $A_0 = 1, A_{-1} = 0$, and similarly for B_j and C_j, with $B_0 = q_0 = 2$, $B_{-1} = 1$. On a single triangular lattice, the two vectors are orthogonal, with nearly the same length. They form a square with a slight orthorhombic distortion. In the overlap, one recognizes the "square" with approximant of $\sqrt{3} \approx 2/1$, the unit cell made of orthogonal mirror planes with $\sqrt{3} \approx 5/3$. In the gap structure, there is the unit cell with $\sqrt{3} \approx 7/4$.

(2) Other coincidence points do not lie on a symmetry axis. They are represented on a single triangular lattice by the two lattice vectors $\boldsymbol{b_1} = (1/2)p\boldsymbol{i} + (\sqrt{3}/2)q\boldsymbol{j}$ and $\boldsymbol{b_2} = (1/2)m\boldsymbol{i} - (\sqrt{3}/2)n\boldsymbol{j}$, with p and q, resp. m and n, integers of same parity. The two vectors have the same length $|\boldsymbol{b_1}| = |\boldsymbol{b_2}|$, and are nearly orthogonal. They form a rhombus that is almost a square. Its diagonals $\boldsymbol{b_1} + \boldsymbol{b_2}$ and $\boldsymbol{b_2} - \boldsymbol{b_1}$ lie on me axes of symmetry of the triangular lattice, thus

$$p - m = q + n, \tag{8.1}$$

$$3(n - q) = p + m. \tag{8.2}$$

Moreover,

$$\boldsymbol{b_1} \cdot \boldsymbol{b_2} = (3q^2 - m^2)/2. \tag{8.3}$$

Note that $m > n$, since $m = n$ would imply $q = 0$. (The two vectors form also an equilateral triangle, so that $\alpha + \beta = \pi/6$, where α and β are the angles between the $\boldsymbol{b}$'s and the symmetry axes. For example, $\tan\beta = m/(n\sqrt{3})$, so that $m < n$.)

This yields two approximants for $\sqrt{3}$, p/n and m/q, with $q < m < n < p$. The solution of (8.1) and (8.2) is,[11]

[10] $1+\sqrt{3} = [\overline{2,1}] = [q_0, q_1, q_2, q_3, \ldots]$, where $q_j = 2$ for j even, $q_j = 1$ for j odd (see footnote 2).

[11] Proof (by induction): $(p-m) - (n+q) = q_{j+2}B_{j+1} - 2A_{j+2} = q_{j+2}[q_{j+1}B_j - 2A_{j+1}] + [q_{j+2}B_{j-1} - 2A_j]$, with $q_{j+2} = q_j$ for $\sqrt{3}$. The two $[\ldots] = 0$ by induction, thus $q_{j+2}B_{j+1} = 2A_{j+2}$. Similarly for (8.2): $3(n-q) - (p+m) = 3q_{j+2}A_{j+1} - 2C_{j+2} + q_{j+2}C_{j+1} = 2q_{j+2}A_{j+1} - 2C_{j+2} + q_{j+}B_{j+1} = 2q_{j+2}A_{j+1} - 2(C_{j+2} - A_{j+2})$, using the result in the proof of (8.1). Then, $q_{j+2}A_{j+1} - (C_{j+2} - A_{j+2}) = q_{j+2}[q_{j+1}A_j - (C_{j+1} - A_{j+1})] + [q_{j+2}A_j - (C_{j+1} - A_{j+1})]$, with $q_{j+2} = q_j$ for $\sqrt{3}$. Once again, the two $[\ldots]$ vanish by induction. Note that we have only used the fact that $q_{j+2} = q_j$, i.e. a continuous fraction expansion of period 2.

$$q = A_j,\ m = C_j,\ n = A_{j+2},\ p = C_{j+2}. \tag{8.4}$$

The smallest solution ($j = 1$) is $q = 1$, $m = 2$, $n = 4$, $p = 7$. It corresponds to the square of the gap structure. The intermediate convergent $q = 2$, $m = 3$, $n = 7$, $p = 12$, which corresponds to the unit cell of overlap structure, yields the same approximants for $\sqrt{3}$.

With $q = A_j$, $m = C_j$, the two vectors are nearly orthogonal. The scalar product $-2\boldsymbol{b_1} \cdot \boldsymbol{b_2} = (m^2 - 3q^2) = 1$ for j odd, $= -2$ for j even, regardless of the length of the vectors. Indeed, $m^2 - 3q^2 = 1$ is known as Pell's equation [5]. Pell's equation has infinitely many solutions C_j/A_j (j odd). For intermediate convergents (defined, for j even, as $A_j^{(1)} = A_{j-1} + A_{j-2}$, $A_j = 2A_{j-1} + A_{j-2} = A_j^{(1)} + A_{j-1}$), Pell's equation is $m^2 - 3q^2 = -3$, i.e. $q^2 - 3(m/3)^2 = 1$, and $q = A_j^{(1)} = C_{j-1}$, $m = C_j^{(1)} = 3A_{j-1}$, yielding the same approximant for $\sqrt{3} \approx C_{j-1}/A_{j-1} = 3A_{j-1}/C_{j-1}$, thus $\sqrt{3} \approx C_{j-1}/A_{j-1}$ as that for $j-1$ odd.

Okuyama et al. [1] mention an average distortion of 1.019 in their crystal of $[Gly - Pro - Pro]_{10}$, which is consistent with these figures.

It is possible to represent the cross-section of the collagen molecule (triple helix) as a trefoil of three, close-packed hexagons drawn on the underlying triangular lattice [3, 10]. Contact between trefoils is through an edge of the triangular lattice in the overlap, and through a vertex in the gap structure. The trefoil rotates as it goes along the molecule. In the overlap, the smaller rhombi have four trefoils filling space without any vacant space. The larger rhombi have a hexagonal hole between the four trefoils, as do the "squares". The distances between trefoil centres are 3 and $2\sqrt{3}$ edges of the underlying triangular lattice. When the trefoils rotate, they push each other apart to reach a single maximum distance in the gap. With one out of every five trefoils missing, the area occupied remains constant. Further rotation of the trefoils leads to an overlap structure, rotated from the first, with the smaller rhombi replaced by larger ones, and vice versa.

8.9 Twist Grain Boundary Overlap–(Gap)–Overlap

Figures 8.8 and 8.9 show how the periodic stack overlap–gap–overlap–... can be constructed, and that the boundaries between gap and overlap structures are twist grain boundaries, associated with a coincidence site lattice.

The precise value of the rotation angle θ' at the twist grain boundary between two overlap structures (separated by a gap) is twice $\theta' = \tan^{-1}(\sqrt{3}/6) + \tan^{-1}(\sqrt{3}/9)$ or $\theta' = \tan^{-1}(5\sqrt{3}/17)$ whose numerical value is $26.996° \simeq 27°$. Namely the rotation is $3\pi/10$. The tenfold symmetry is visible in the diffraction pattern of a single overlap structure. It corresponds to a true coincidence site lattice.

In practice, we place the vertex at the origin of the rotated structure anywhere within the Voronoi cell of one of the five vertices of the unit cell of the overlap structure. The gap structure, rotated around one of its vertices by 27°, has vertices in four out of five Voronoi cells of the original overlap structure (as indicated schematically in Fig. 8.5b). The next overlap, rotated by a further 27°, has its vertices inside all five Voronoi cells of the original structure. The vertex superposition is only topological; the molecule at the origin of the structures is only approximatively straight.

The operation is repeated, starting now with the rotated overlap structure. The primitive cell is translated to have its basis in the same orientation. But the position of the smaller and larger rhombi is now exchanged, as was suggested in the trefoil model of Bouligand. This is another effect of the rhombic distortion of the perfect square–triangle structure. Accordingly, the rotation angle of the TGB is now twice $\tan^{-1}(\sqrt{3}/7)+\tan^{-1}(\sqrt{3}/5) = \tan^{-1}(3\sqrt{3}/8) \approx 33° = \pi/3-3\pi/20$, the complementary – to $2\pi/3$ – of $3\pi/10$. Hence, the apparent tenfold rotation, superposed on the sixfold triangular symmetry, noticeable in the diffraction patterns. The apparent contradiction between microfibrils with pentagonal cross-section [11] and quasi-hexagonal alignment of triple helices [12] is hereby resolved: the collagen fibril exhibits both rotations symmetries (Figs. 8.6, 8.8 and 8.9), enforced by the orthorhombic distortion of the square–triangle pattern on a triangular lattice.

References

1. Ke. Okuyama, Ka. Okuyama, S. Arnott, M. Takayanagi, M. Kakudo, J. Mol. Biol. **152**, 427 (1981)
2. A. Rich and F.R.C. Crick, J. Mol. Biol. **3**, 483 (1961)
3. Y. Bouligand, L'assemblage compact des triples hélices de collagène, JMC5, poster2108 (Soc. Franç. Phys., Orléans, 1997).
4. J.F. Sadoc and N. Rivier, Eur. Phys. J. **B 12**, 309 (1999)
5. H. Davenport, Hutchinson, *The Higher Arithmetic* (London, 1952)
6. J.F. Sadoc and R. Mosseri, *Geometrical Frustration*, (Cambridge University Press, Cambridge, 1999)
7. N.S. Manton, Commun. Math. Phys. **113**, 341 (1987)
8. J.A. Hodge and A.J. Petruska, Recent studies with the electron microscope on ordered aggregates of the tropocollagen molecule, in *Aspects of Protein Structure*, ed. by G.N. Ramachandran (Academic, London, 1963), pp. 289–300
9. L. Navailles, B. Pansu, L. Gorre-Talini and H.T. Nguyen, Phys. Rev. Lett. **81**, 4168 (1998)
10. Y. Bouligand, Two superposed fibre diagrams in X-ray diffraction patterns of collagen tendons: The square-lozenge model. Unpublished (2001)
11. J.W. Smith, Nature **219**, 157 (1968)
12. D.J.S. Hulmes and A. Miller, Nature **282**, 878 (1979)

9

Euler Characteristic, Dehn–Sommerville Characteristics, and Their Applications

V.M. Buchstaber

Summary. In this chapter we present several classical results centring on the notion of Euler characteristic of simplicial complexes and manifolds. We also consider some results that are not discussed in the textbooks of topology. These results are concerned with construction of certain combinatorial invariants of manifold triangulations, which we call the *Dehn–Sommerville characteristics.*

9.1 Introduction

In May 2002 the author read a mini-course of lectures in algebraic topology at the Max Planck Institute, Dresden. As the audience consisted mostly of physicists and biologists, the course aimed at introducing several fundamental concepts, requiring only basic mathematical knowledge. The notes from the lecture course have grown up into this text.

We give several classical results centring on the notion of Euler characteristic of simplicial complexes and manifolds. The chapter also contains some results that have not yet found their way into algebraic topology textbooks, but are of considerable interest due to their applications in several fields, including discrete mathematical physics. These results are connected with construction and applications of certain combinatorial invariants of manifold triangulations, which we call the *Dehn–Sommerville characteristics.*

All the key properties are illustrated on the triangulations of two-dimensional surfaces. Detailed proofs and further developments for most of the results of this study can be found in [1, 2].

9.2 Simplicial Complexes and Maps

Denote $\mathbb{R}^n$ as an n-dim Eucledian space. An *n-dim simplex* σ^n is the convex hull in $\mathbb{R}^n$ of any $(n+1)$ points $\alpha_0, \dots, \alpha_n$ not contained in an $(n-1)$-dim hyperplane.

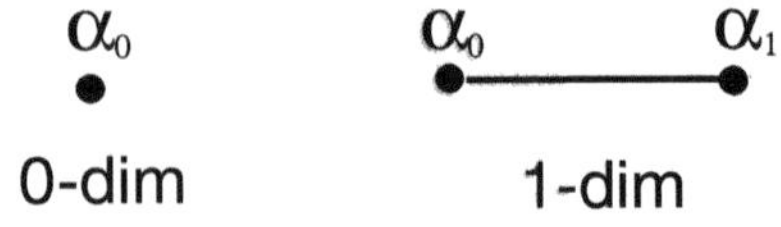

Fig. 9.1. Simplex

A point $x \in \sigma^n$ can be written in *barycentric coordinates* as

$$x = \sum_{j=0}^{n} x^j \alpha_j, \qquad \sum_{j=0}^{n} x^j = 1, \quad x^j \geq 0.$$

A *face* of σ^n is the simplex determined by a subset of vertices $\alpha_0, \dots, \alpha_n$. The empty subset of vertices determined the empty face.

Example 1. A face of dim $(n-1)$ is given by $\sigma_j^{n-1} = (\alpha_0, \dots, \widehat{\alpha_j}, \dots, \alpha_n)$.

A finite *simplicial complex* K is a finite collection of simplices satisfying the following two properties:

1. Each face of a simplex from the collection belongs to the collection
2. The intersection of any two simplices from the collection is a face of each

Example 2. The boundary of an n-dim simplex σ^n is the union $\cup_j \sigma_j^{n-1}$ of its $(n-1)$-dim facets, together with all their faces. This is an $(n-1)$-dim simplicial complex, the *standard* simplicial subdivision of the sphere S^{n-1}.

A *map of simplices* $\sigma_1^n \to \sigma_2^m$ is a map from the vertices of σ_1^n to the vertices of σ_2^m extended linear to the whole of σ_1^n. A *simplicial map* $f : K_1 \to K_2$ of complexes is a map whose restriction to each simplex is a map of simplices. Therefore, a simplicial map is determined by the images $f(\alpha_j) = \beta_k$, where $\{\alpha_j\}$ and $\{\beta_k\}$ are the sets of vertices of K_1 and K_2.

Example 3. Let K be any simplicial complex on the vertex set $\{v_0, \dots, v_{m-1}\}$, and σ^{m-1} the standard simplex on the vertices $\{\alpha_0, \dots, \alpha_{m-1}\}$. Then there exists a canonical simplicial map (inclusion)

$$f : K \hookrightarrow \sigma^{m-1}$$

determined by $f(v_j) = \alpha_j$.

Given two simplicial complexes K_1 and K_2, we say that K_2 is a subdivision of K_1 if each simplex of K_1 is a union of finitely many simplices of K_2 and the simplices of K_2 are contained linearly in the simplices of K_1.

The barycentric subdivision K' provides a standard way to subdivide any simplicial complex K.

The barycentric subdivision may be defined inductively: To subdivide an n-simplex σ^n we first barycentrically subdivide the faces of σ^n, then introduce yet another vertex $\alpha \in \sigma^n$ in the centre of σ^n, and add new simplices of the form $(\beta_0, \dots, \beta_k, \alpha)$.

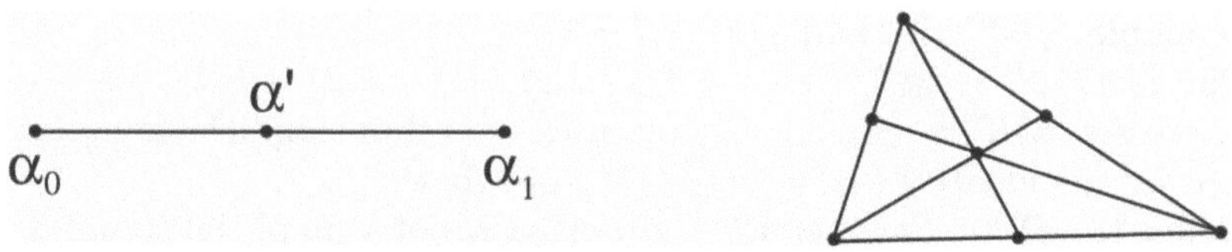

Fig. 9.2. Barycentric subdivision

Example 4. Let K be an $(n-1)$-dim simplicial complex on the vertex set $\{v_0, \ldots, v_{m-1}\}$, and σ^{n-1} the standard simplex on the vertices $\{\alpha_0, \ldots, \alpha_{n-1}\}$. Then there exists a canonical simplicial map

$$\varphi : K' \hookrightarrow \sigma^{n-1}$$

determined by $\varphi(\omega) = \alpha_k, \;\; 1 \leq k \leq n$, where $\omega = (v_{i_1}, \ldots, v_{i_k})$ is a simplex of K.

The map φ from the previous example belongs to a very special class of simplicial maps, the so-called *branched combinatorial coverings.*

A map $p : K_1 \to K_2$ between simplicial complexes K_1 and K_2 is called a *branched combinatorial covering* if:

1. For any relatively open simplex $\mathring{\tau} \in K_2$ the preimage $p^{-1}(\mathring{\tau})$ is a finite non-empty disjoint union of relatively open simplices $\mathring{\omega}_i(\tau)$;
2. The map $p : \mathring{\omega}_i(\tau) \to \mathring{\tau}$ is a homeomorphism for all i.

Two simplicial complexes K_1 and K_2 are said to be *combinatorially equivalent* if there exists a simplicial complex K isomorphic to a subdivision of each of them. The combinatorial neighbourhood of a simplex $\tau \in K$ is the subcomplex consisting of all simplices, together with their boundaries having the simplex τ as a face. A simplicial complex K is called an *n-dim piecewise linear (PL-) manifold* if after application of a sequence of barycentric subdivisions the combinatorial neighbourhood of each simplex becomes a complex combinatorially equivalent to σ^n.

9.3 Euler Characteristic and Dehn–Sommerville Characteristics

The f-vector of an $(n-1)$-dim simplicial complex K is given by

$$\boldsymbol{f}(K) = (f_0, f_1, \ldots, f_{n-1}),$$

where f_i is the number of i-dim simplices of K^{n-1}.

The *Euler characteristic* of K^{n-1} is

$$\chi(K^{n-1}) = f_0 - f_1 + \cdots + (-1)^{n-1} f_{n-1}.$$

For example, $\chi(\sigma^n) = 1$ and $\chi(S^{n-1}) = 1 + (-1)^{n-1}$.

Put $f(t) = t^n + f_0 t^{n-1} + \cdots + f_{n-1}$ and $h(t) = h_0 t^n + h_1 t^{n-1} + \cdots + h_n$. The *h-vector* $\boldsymbol{h}(K) = (h_0, \dots, h_n)$ of an $(n-1)$-dim simplicial complex K is defined by the identity $h(t) = f(t-1)$. Note that $h_0 = 1$.

Define the *Dehn–Sommerville characteristics* of a simplicial complex K by the formula:

$$\mathrm{DS}_i(K) = (-1)^{n-1}(h_{n-i} - h_i), \qquad i = 0, \dots, n.$$

The numbers DS_i are obviously combinatorial invariants of a simplicial complex K. We have

$$\mathrm{DS}_0(K) = \chi(K) - \chi(S^{n-1}).$$

Two maps $f_1, f_2 : X \to Y$ are called *homotopic* if there is a continuous map $F : X \times I \longrightarrow Y$ (here I is the interval $[0,1]$) such that $F(x,0) = f_1(x)$ and $F(x,1) = f_2(x)$ for all $x \in X$.

Fix a basepoint $\mathrm{pt} \in X$. Homotopy classes of maps $\varphi : I \to X$ such that $\varphi(0) = \varphi(1) = \mathrm{pt}$ form a group called the *fundamental group* of X and denoted $\pi_1(X)$.

A continuous map $f : X \to Y$ is called a *homotopy equivalence* if there is a map $g : Y \to X$ such that the two composites $g \circ f : X \to X$ and $f \circ g : Y \to Y$ are homotopic to the identity maps id_X and id_Y, respectively.

A characteristic $a(X)$ of a space X is called a *homotopy invariant* if $a(X) = a(Y)$ whenever there is a homotopy equivalence $f : X \to Y$.

The Euler characteristic $\chi(X)$ is a homotopy invariant, and therefore so is $\mathrm{DS}_0(K)$.

For $i > 0$ the characteristic $\mathrm{DS}_i(K)$ is not homotopy invariant in general. Remarkably, it becomes a homotopy invariant if we restrict to triangulated manifolds. More precisely, for any triangulated manifold K^{n-1} the following *generalised Dehn–Sommerville relations* hold:

$$\mathrm{DS}_i(K) = (-1)^i \left(\chi(K^{n-1}) - \chi(S^{n-1})\right) \binom{n}{i}, \quad i = 0, 1, \dots, n.$$

If K is a simplicial subdivision of the sphere S^{n-1} we obtain

$$\mathrm{DS}_i(K) = 0.$$

The Dehn–Sommerville relations are the most general linear equations satisfied by the f-vectors of triangulated spheres.

Example 5. Let K be a simplicial subdivision of the sphere S^2. Then it follows from the Dehn–Sommerville relations that

$$\boldsymbol{f}(K) = \left(f_0, 3(f_0 - 2), 2(f_0 - 2)\right).$$

9.4 Homology Groups and Characteristic Classes

Given a *simplicial complex* K, we fix an order of its vertices $\alpha_0, \alpha_1, \dots, \alpha_m$. Then an r-dim simplex of K can be written as $[\alpha_{i_0}, \alpha_{i_1}, \dots, \alpha_{i_r}]$, $i_0 < \dots < i_r$. This fixes a canonical *orientation* on it.

Suppose that we are also given an *abelian group* G with operation "+". A k-dimensional *chain* of K with coefficients in G is a finite linear combination of distinct k-simplices of K of the form

$$c_k = \sum_i g_i \sigma_i, \qquad g_i \in G.$$

Chains of K of dimension k form an abelian group $C_k(K)$ with the sum of two chains c_k and $c'_k = \sum_i g'_i \sigma_i$ given by

$$c_k + c'_k = \sum_i (g_i + g'_i)\sigma_i.$$

The *boundary* of an n-simplex $\sigma^n = [\alpha_0, \dots, \alpha_n]$ is the $(n-1)$-chain

$$\partial \sigma^n = \partial[\alpha_0, \dots, \alpha_n] := \sum_{i=0}^{n} (-1)^i \sigma_{(i)}^{n-1},$$

where $\sigma_{(i)}^{n-1} := [\alpha_0, \dots, \widehat{\alpha_i}, \dots, \alpha_n]$. For example,

$$\begin{aligned} \partial[\alpha_0] &= 0, \\ \partial[\alpha_0, \alpha_1] &= [\alpha_1] - [\alpha_0], \\ \partial[\alpha_0, \alpha_1, \alpha_2] &= [\alpha_1\alpha_2] - [\alpha_0\alpha_2] + [\alpha_0\alpha_1]. \end{aligned}$$

For any n-dim simplex we have

$$\partial\partial[\alpha_0 \dots \alpha_n] = \sum_{i=0}^{n} (-1)^i \partial\sigma_{(i)}^{n-1} = 0.$$

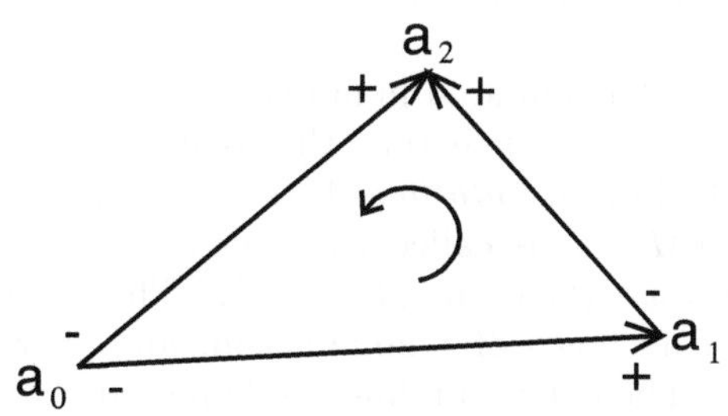

Fig. 9.3. Boundary

The *boundary* of an arbitrary k-chain $c_k = \sum_i g_i \sigma_i$ is then given by

$$\partial c_k := \sum_i g_i \partial \sigma_i.$$

Again, we have $\partial\partial c_k = 0$.

The *k-cycles* of K are those k-chains c_k satisfying $\partial c_k = 0$. They form a subgroup denoted Z_k.

The *boundary* k-cycles are those that are "homologous to zero", i.e. are of the form ∂c_{k+1} for some $(k+1)$-chain c_{k+1}. The subgroup they comprise is denoted by B_k.

We say that two chains c'_k and c''_k are *homologous* if

$$c'_k = c''_k + \partial c_{k+1}$$

for some $(k+1)$-chain c_{k+1} of K. The k-dim *homology* group $H_k(K;G)$ is the quotient of Z_k/B_k.

The fundamental group is not commutative in general, and its abelianisation is the first homology group:

$$H_1(K;\mathbb{Z}) \cong \pi_1(K)/\big[\pi_1(K),\pi_1(K)\big],$$

where $[\pi_1(K),\pi_1(K)]$ is the *commutator subgroup* of the fundamental group.

When $G = \mathbb{R}$ is the group of real numbers, the group $H_k(K;G)$ is a real vector space. By definition, the kth *Betti number* b_k of M is $\dim H_k(K;\mathbb{R})$.

A theorem of Poincaré states that

$$\chi(K) = \sum_{i\geq 0}(-1)^i f_i = \sum_{i\geq 0}(-1)^i b_i.$$

If M^n is a closed and connected manifold admitting a finite triangulation, then

$$H_n(M^n;\mathbb{Z}_2) = \mathbb{Z}_2,$$

where $\mathbb{Z}_2$ is the 2-element group of residues modulo 2. The generator of $H_n(M^n;\mathbb{Z}_2)$ is given by the homology class of the chain $\sum_i \sigma_i^n$, where the sum is taken over all simplices of K. This generator is called the *$\mathbb{Z}_2$-fundamental class* of M^n.

A closed connected triangulated manifold M^n is called *orientable* if there is a choice of signs $\varepsilon_i = \pm 1$ such that the n-chain $\sum_i \varepsilon_i \sigma_i$ is a cycle. This choice of signs is called an *orientation* of M^n, and the homology class of the cycle $\sum_i \varepsilon_i \sigma_i$ in $H_n(M^n;\mathbb{Z})$ is called the *fundamental class* of the oriented manifold M^n. It generates the group $H_n(M^n;\mathbb{Z})$. The property of orientability does not depend on a choice of triangulation, and there are two different orientations in total. The corresponding fundamental classes differ by a sign.

For every closed connected orientable triangulated manifold M^n we have $H_n(M^n;G) = G$ for all G. If M^n is non-orientable then

$$H_n(M^n;\mathbb{Z}) = 0, \qquad H_n(M^n;\mathbb{Z}_2) = \mathbb{Z}_2.$$

A simplicial map $f : K_1 \to K_2$ induces group homomorphisms

$$f_* : H_k(K_1; G) \to H_k(K_2; G), \qquad k = 0, 1, \ldots.$$

In particular, for a simplicial map $f : M_1^n \to M_2^n$ between two oriented manifolds we have $f_*[M_1^n] = m[M_2^n]$, where m is called the *degree* of f (the same is true for non-orientable manifolds if we work with $\mathbb{Z}_2$-homology).

Consider the barycentric subdivision M' of a manifold $M = M^n$ and define

$$w_k = \sum_i \omega_i^k,$$

where the sum is taken over all k-dim simplices ω_i^k of M'. It is remarkable that for any k the chain w_k is a cycle with coefficients $\mathbb{Z}_2$. The homology class $[w_k]$ is a homotopy invariant of M^n, called its kth *homology Stiefel–Whitney* class. In particular, $[w_n] = [M^n]$ and $w_0 = \chi(M^n) \bmod 2$.

9.5 Classification of 2-Manifolds

Any closed orientable surface is homeomorphic to a sphere with g handles. The integer g is called the *genus*. A model of a closed orientable surface of genus g is given by a non-singular *hyperelliptic curve*:

$$V = \{(x, y) \in \mathbb{C}^2 : y^2 = \mathcal{F}(x)\},$$

where

$$\mathcal{F}(x) = 4x^{2g+1} + \lambda_{2g}x^{2g} + \cdots + \lambda_1 x + \lambda_0$$

is a polynomial with all distinct roots.

A closed orientable surface M_g^2 of genus g can be obtained by a suitable identification of edges in a $4g$-gon.

The fundamental group $\pi_1(M_g^2)$ has $2g$ generators $a_1, \ldots, a_g, b_1, \ldots, b_g$ with a single defining relation

$$a_1 b_1 a_1^{-1} b_1^{-1} \cdots a_g b_g a_g^{-1} b_g^{-1} = 1,$$

coming from the identification of edges.

A closed non-orientable surface can be obtained by a suitable identification of edges in a $4g$- or $(4g+2)$-gon. Therefore, there are two families of non-orientable closed surfaces, $N_{g,1}^2$ and $N_{g,2}^2$.

The corresponding relations in the fundamental group are:

$$N_{g,1}^2\colon \left(\prod_{i=1}^{g-1} a_i b_i a_i^{-1} b_i^{-1}\right) a_g b_g a_g^{-1} b_g = 1,$$
$$N_{g,2}^2\colon \left(\prod_{i=1}^{g} a_i b_i a_i^{-1} b_i^{-1}\right) c^2 = 1.$$

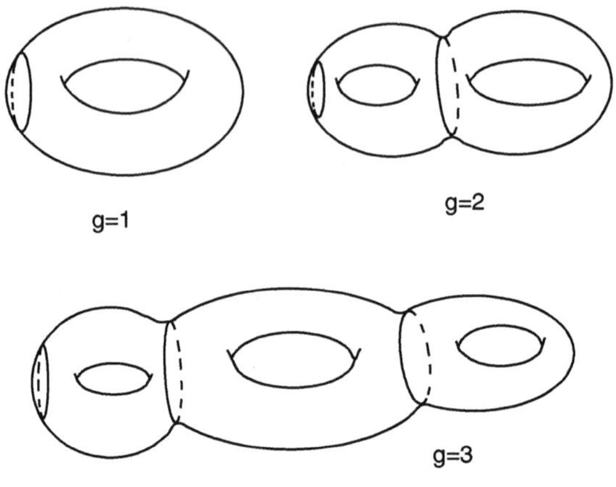

Fig. 9.4. Sphere with g handles

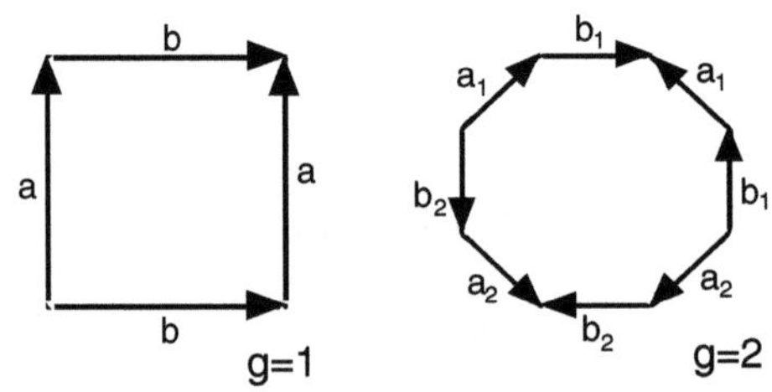

Fig. 9.5. M_g^2: $4g$ edges

An orientable closed surface M_g^2 (a sphere with g handles) can be smoothly embedded in $\mathbb{R}^3$ as the boundary of a 3-dim body.

A model of the 3-body with boundary M_g^2 can be obtained by taking the small closed smooth neighbourhood of the wedge of g circles in $\mathbb{R}^3$.

Any non-orientable closed surface can be obtained as follows. Take a sphere S^2, remove μ disjoint open disks D^2, and identify the diametrically opposite points on the boundary of each hole. This is equivalent to filling all the μ holes by Möbius bands (crosscaps). Denote the resulting surfaces by M_μ^2, $\mu = 1, 2, \ldots$.

The diffeomorphism classes of connected closed manifolds M^2 form a *commutative semigroup* with respect to the connected sum operation #. This semigroup has two generators a (the torus T^2) and b (the projective plane $\mathbb{R}P^2$) with a single defining relation

$$a\#b = b\#b\#b.$$

Every connected closed smooth manifold M^2 admits a finite triangulation, i.e. can be subdivided by means of smooth curves into finitely many smooth triangles in such a way that any two triangles either do not intersect, have a single common vertex (0-face), or share a single common edge (1-dim face).

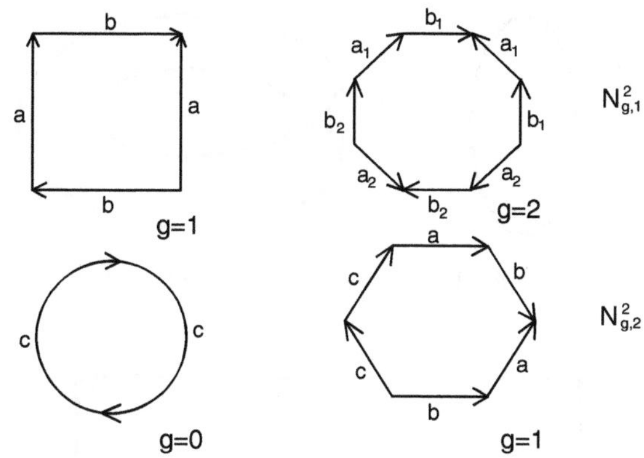

Fig. 9.6. $N^2_{g,1}$ and $N^2_{g,2}$

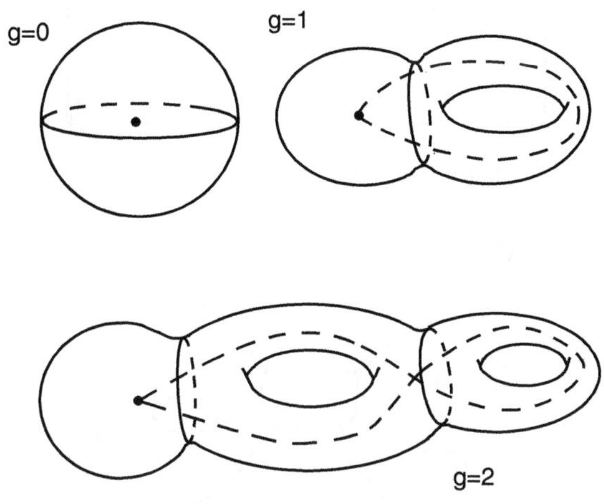

Fig. 9.7. 3-Dim body

The first homology of 2-dim surfaces is given by

$$H_1(M_g^2;\mathbb{Z}) = \underbrace{\mathbb{Z}\oplus\cdots\oplus\mathbb{Z}}_{2g};$$

$$H_1(M_\mu^2;\mathbb{Z}) = \underbrace{\mathbb{Z}\oplus\cdots\oplus\mathbb{Z}}_{\mu-1}\oplus\mathbb{Z}_2.$$

It follows that

$$\chi(M_g^2) = 2-2g, \quad \chi(M_\mu^2) = 2-\mu.$$

Any closed non-orientable surface M_μ^2 can be obtained from the orientable surface M_g^2 with $g=\mu-1$ by taking the orbit space of a certain involution.

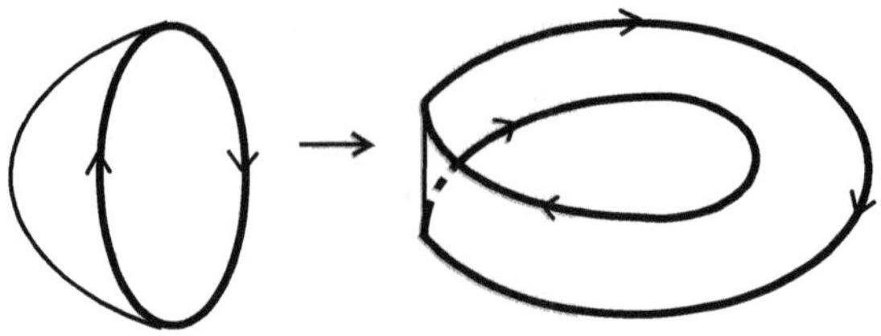

Fig. 9.8. $M^2_{\mu=1}$: real projective plane $\mathbb{R}P^2$

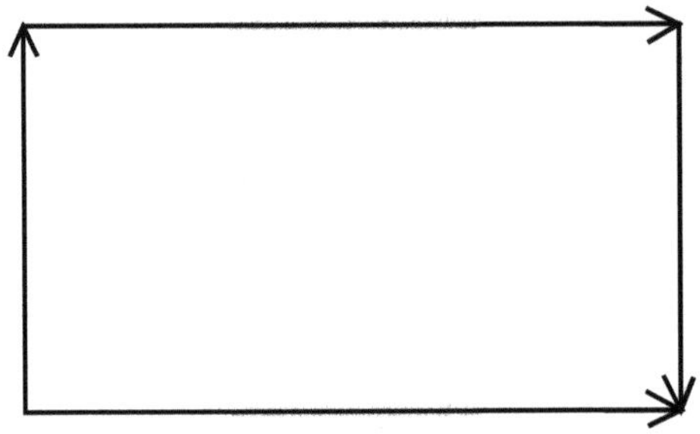

Fig. 9.9. $M^2_{\mu=2}$: Klein bottle $\mathbb{K}^2$

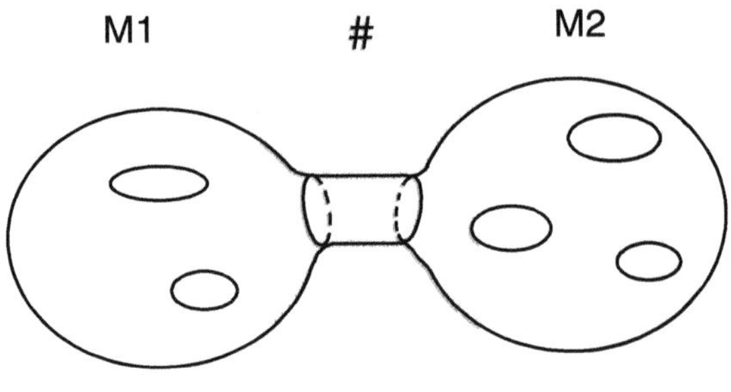

Fig. 9.10. Connected sum

9.6 Minimal and Neighbourly Triangulations

A triangulation $\mathcal{T}_*$ of a manifold M^n is called *minimal* if $f_0(\mathcal{T}_*) \leqslant f_0(\mathcal{T})$ for any triangulation $\mathcal{T}$ of M^n, where f_0 denotes the number of vertices.

Now assume that M^2 is a 2-dim triangulated manifold with m vertices. Let $\chi = \chi(M^2)$ be its Euler characteristic. Then using the Dehn–Sommerville equation we obtain

$$\boldsymbol{f}(M^2) = \big(m, 3(m-\chi), 2(m-\chi)\big).$$

For any simplicial complex K on m vertices there exists a simplicial inclusion $K \hookrightarrow \sigma^{m-1}$ (see Example 2). Therefore,

$$f_1(K) \le \binom{m}{2}.$$

For 2-dim triangulations we obtain:

$$6(m-\chi) \le m(m-1).$$

This leads to the following lower bounds for the number of vertices $f_0 = m$ in triangulations of particular 2-dim manifolds, e.g.

$$\text{2-dim torus } T^2\colon \chi(T^2) = 0 \text{ and } m \ge 7;$$
$$\text{Projective plane } \mathbb{R}P^2\colon \chi(\mathbb{R}P^2) = 1 \text{ and } m \ge 6.$$

A triangulation $\mathcal{T}$ of a manifold M^2 is called *neighbourly* if its 1-skeleton is a complete graph (i.e. any two vertices are joined by an edge).

Theorem 1. *Let $\mathcal{T}$ be a minimal triangulation of a closed manifold M^2. Then $\mathcal{T}$ is neighbourly if and only if*

$$6(m-\chi) = m(m-1),$$

where $m = f_0(M^2)$ and $\chi = \chi(M^2)$. In this case:

(a) If M^2 is orientable of genus g, then

$$g \in \{(3q-1)(4q-1),\ (3q+1)(4q+1),\ q(12q-1),\ q(12q+1), \quad q = 0, 1, \ldots\}.$$

(b) If M^2 is non-orientable with μ crosscaps, then

$$\mu \in \{(2q-1)(3q-2),\ (2q-1)(3q-1),\ q(6q-1),\ q(6q+1), \quad q = 0, 1, \ldots\}.$$

Example 6. For $q = 0$ the possible values of g are $1, 0$, and the possible values of μ are $2, 1, 0$. For $q = 1$ we get $g \in \{6, 20, 11, 13\}$ and $\mu \in \{1, 2, 5, 7\}$.

Minimal neighbourly triangulations exist for the sphere S^2 ($g = 0, m = 4$), torus T^2 ($g = 1, m = 7$), and real projective plane $\mathbb{R}P^2$ ($\mu = 1, m = 6$). However for most values of χ there is no minimal neighbourly triangulation (see the previous theorem). For example, minimal triangulations of orientable surfaces of genus 2 to 5 are not neighbourly, and minimal triangulations of non-orientable surfaces with μ crosscaps are not neighbourly for $\mu = 3, 4$.

9.7 Smooth Manifolds

A *smooth n-manifold* M^n is covered by open subsets U_α:

$$M^n = \cup_\alpha U_\alpha.$$

For each U_α there is fixed a homeomorphism $\varphi_\alpha : U_\alpha \to \mathbb{R}^n$ providing the *local coordinates* $x_\alpha^1, \dots, x_\alpha^n$ in U_α. Therefore, there are two sets of coordinates on each intersection $U_\alpha \cap U_\beta$, namely, for every $x \in U_\alpha \cap U_\beta$ we have

$$\varphi_\alpha(x) = (x_\alpha^1, \dots, x_\alpha^n) \quad \text{and} \quad \varphi_\beta(x) = (x_\beta^1, \dots, x_\beta^n).$$

The *coordinate transformation* is given by the set of smooth functions:

$$\begin{aligned} x_\alpha^j &= f_\alpha^j(x_\beta^1, \dots, x_\beta^n), \quad j = 1, \dots, n; \\ x_\beta^k &= g_\beta^k(x_\alpha^1, \dots, x_\alpha^n), \quad k = 1, \dots, n, \end{aligned}$$

where the composition $f \circ g$ is the identity transformation of $\mathbb{R}^n$. A *smooth map* $F : M \to N$ between two manifolds is given by smooth functions in any local coordinate system.

Suppose we are given a curve segment $x = x(\tau) \in M,\ \ a \le \tau \le b$, on a manifold M. The part of the curve belonging to a coordinate region U_α is described by the set of *parametric equations*

$$x_\alpha^j = x_\alpha^j(\tau), \quad j = 1, \dots, n.$$

The *velocity* (or *tangent*) *vector* at a point $x = x(\tau)$ is given by

$$\dot{x} = (\dot{x}_\alpha^1, \dots, \dot{x}_\alpha^n).$$

In the intersection $U_\alpha \cap U_\beta$ we can write the parametric equations $x_\alpha(\tau)$ and $x_\beta(\tau)$ in the two coordinate systems. Using the coordinate transformation formulae we obtain

$$x_\alpha^j(\tau) = f_\alpha^j\left(x_\beta^1(\tau), \dots, x_\beta^n(\tau)\right).$$

Therefore,

$$\dot{x}_\alpha^j = \sum_k \left(\frac{\partial f_\alpha^j}{\partial x_\beta^k}\right) \dot{x}_\beta^k.$$

An n-dim manifold M^n is called *orientable* if there is a coordinate covering $M^n = \cup_\alpha U_\alpha$ such that the coordinate transformations satisfy

$$\det\left(\frac{\partial f_\alpha^j}{\partial x_\beta^k}\right) > 0$$

for all $x \in U_\alpha \cap U_\beta$.

A smooth manifold M^n that can be smoothly embedded in $\mathbb{R}^{n+1}$ is orientable. It follows that a 2-dim manifold is embeddable in $\mathbb{R}^3$ if and only if it is orientable.

A *tangent* vector to an n-dim manifold M^n is by definition the velocity vector of some smooth curve. In any system of local coordinates x_α^j the tangent vector at a point $x \in M$ can be written as an n-tuple (ξ_α^j). The two n-tuples corresponding to different coordinate systems are connected by the formula:

$$\xi_\alpha^j = \sum_{k=1}^{n} \left(\frac{\partial f_\alpha^j}{\partial x_\beta^k} \right) \xi_\beta^k.$$

The set of all tangent vectors to M at a point x forms an n-dim linear space $T_x = T_x M$, called the *tangent space* to M at x. A smooth map $F : M \to N$ induces a linear map

$$F_* : T_x \to T_{F(x)}$$

at any $x \in M$.

Denote by TM the *tangent bundle* of M. It consists of pairs (x, η), where $x \in M$ and η is a tangent vector to M at x. A (tangent) *vector field* on M is a smooth map $\xi : M \to TM$ such that $\xi(x) \in T_x M$. It follows that ξ is an embedding; we denote its image by $M(\xi)$. We identify the manifold M with its image $M(0) \subset TM$ under zero vector field. Note that $\dim TM = 2n$ if $\dim M = n$.

A vector field ξ on M is said to be of *general position* if $M(\xi)$ and $M = M(0)$ intersect transversally, that is, in a finite number of points. These points are called *singular*. All singular points x_j of a general positioned vector field on an oriented manifold M are *non-degenerate* in the sense that

$$\left(\det \frac{\partial \xi^k}{\partial x^j} \right)_{x_j} \neq 0.$$

The *index* of a singular point x_j is $\left(\det \frac{\partial \xi^k}{\partial x^j} \right)_{x_j}$.

Theorem 2 (Hopf–Poincaré). *The Euler characteristic of a closed oriented manifold M^n equals the sum of indices of singular points of any general positioned vector field ξ. In particular, the latter sum does not depend on a vector field.*

An important case of vector fields arises from a smooth function f on M with only non-degenerate critical points x_j. (A point $x \in M$ is called *critical* if $(df)_x = 0$; a critical point is *non-degenerate* if $\det(d^2 f)_x \neq 0$.) Denote by $i(x)$ the number of negative squares in the canonical representation of the quadratic form $(d^2 f)_x$. Then the sum

$$\sum_{j=1}^{m} (-1)^{i(x_j)},$$

over the critical points $x_1, \dots, x_m$ of f equals $\chi(M)$; in particular, it is independent of f.

An (autonomous) *dynamic system* on a manifold M is a smooth vector field ξ on M. In terms of the local coordinates $\{x^j_\alpha\}$ on M, a dynamical system ξ gives rise to the system of (autonomous) *ordinary differential equations*

$$\dot{x}^j_\alpha = \xi^j(x^1_\alpha, \ldots, x^n_\alpha).$$

The solutions of this system are called the *integral curves* or *integral trajectories* of the dynamical system. Therefore, an integral trajectory is a curve $\gamma(t)$ on M whose velocity vector $\dot{\gamma}(t)$ coincides with $\xi(\gamma(t))$ for all t.

Theorem 3. *A closed connected smooth manifold M admits a non-vanishing tangent vector field ξ if and only if its Euler characteristic equals zero.*

Corollary 1.
(a) The torus T^2 is the only orientable surface admitting a non-vanishing tangent vector field.
(b) The Klein bottle $\mathbb{K}^2$ is the only non-orientable surface with a non-vanishing vector field.

Acknowledgements

I cordially thank Natalia Dobrinskaya, Taras Panov, and my wife Galina for their great help in preparing these notes for publication.

References

1. Victor M. Buchstaber, Taras E. Panov, *Torus Actions and Their Applications in Topology and Combinatorics*, University Lecture, vol. 24 (American Mathematical Society, Providence, RI, 2002)
2. Sergei P. Novikov, *Topology I*, Encyclopaedia Mathematical Science, vol. 12 (Springer, Berlin Heidelberg New York, 1996)

10

Hopf Fibration and Its Applications

M. Monastyrsky

Summary. In this chapter we deal with Hopf fibration – one of the key examples in the topology of manifolds – and vividly illustrate the power and diversity of applications of topology. We especially point out less familiar applications of Hopf fibrations. For the sake of volume limit we omit the proofs and refer to the appropriate literature.

10.1 Classical Hopf Fibration

Hopf in his celebrated paper "über die Abbildungen der 3-Sphare auf die Kugelfleche," Math. Ann. **104**, 637–665, (1931) studied the space of homotopically nontrivial mappings of spheres:

$$S^3 \to S^2. \tag{10.1}$$

In modern language it is a group $\pi_3(S^2)$.

Later, more general mappings

$$S^{2n-1} \to S^n \tag{10.2}$$

came to be called *general Hopf fibrations.*

10.1.1 Constructing the Hopf Fibrations

We first show that

$$\pi_3(S^2) = \mathbb{Z}. \tag{10.3}$$

The proof follows immediately from the exactness of the sequence of homotopy groups

$$\pi_i(S^1) \to \pi_i(S^3) \to \pi_i(S^2) \to \pi_{i-1}(S^1). \tag{10.4}$$

For $i \geq 3$ we have

$$0 \to \pi_i(S^3) \to \pi_i(S^2) \to 0.$$

In particular $\pi_3(S^2) = \mathbb{Z}$.

The homotopy classes of maps $S^3 \to S^2$ are characterized by the group of integers $\mathbb{Z}$. What is their geometrical meaning?

Let us consider a special map $p : S^3 \to S^2$, which is called *Hopf fibration* (Fig. 10.1). We represent S^3 as a pair of complex numbers (z_1, z_2) with the point $z_1/z_2 = w$ (w is a point in the complex plane $\mathbb{C}$). The map $(z_1, z_2) \mapsto w$ is extended to the completion of $\mathbb{C}$ with the point $z_2 = 0$ at infinity. We thus obtained a map $S^3 \to S^2 \sim \overline{\mathbb{C}} = \mathbb{C} \cup \infty = \mathbb{CP}(1)$ – Riemann sphere. It is obvious that under this map the points $(\exp(\mathrm{i}\varphi)z_1, \exp(\mathrm{i}\varphi)z_2)$ are sent into the same point w. Therefore the fiber of the bundle $S^3 \to S^2$ is the set of points $\lambda = \exp(\mathrm{i}\varphi) \sim S^1$. We obtained a fiber map $p : S^3 \to S^2$ with the fiber S^1. It is easy to see that p is not trivial, i.e., not equivalent to the direct product

$$S^3 \neq S^2 \times S^1.$$

The simplest way to see this is to calculate the homotopy groups (e.g., π_1 or π_2) of both sides, namely,

$$\text{(a)}\ \pi_2(S^3) = 0,\ \pi_2(S^2 \times S^1) = \mathbb{Z}, \quad \text{(b)}\ \pi_1(S^3) = 0,\ \pi_1(S^2 \times S^1) = \mathbb{Z}.$$

The geometrical meaning of that any circle in S^3 can be contracted to a point but not in $S^1 \times S^2$.

It is possible to prove the following proposition:

Proposition 1. *The set of classes of homotopy maps $S^3 \to S^2$ is the composition of homotopy maps $f : S^3 \to S^3$ and the Hopf fibration $p : S^3 \to S^2$.*

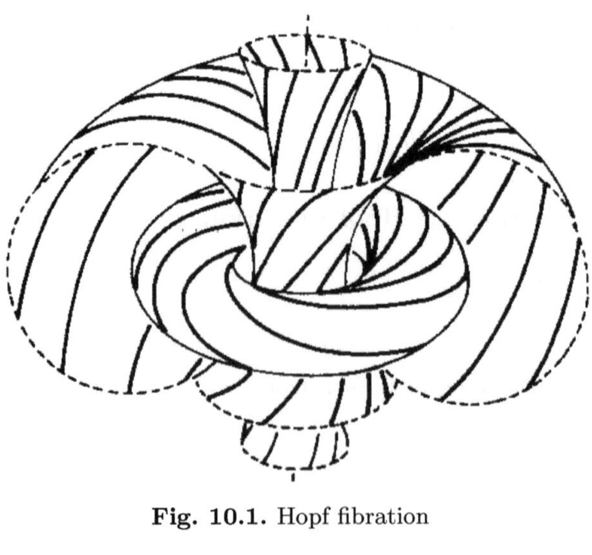

Fig. 10.1. Hopf fibration

(See the proof in [1] or [2].)

We now investigate different properties of Hopf fibrations.

10.1.2 Linking Numbers

Let f be a smooth map $S^3 \to S^2$, y_0, y_1 two regular points in S^2 and M_0 and M_1 the preimages of Y_0 and y_1 equal to $f^{-1}(y_0)$, $f^{-1}(y_1)$, respectively. We define $H(f) = \{M_0, M_1\}$ to be the linking number for the inverse images. It follows from the definition of regularity that $f^{-1}(y_I) \sim S^1$.

Definition 1. *The linking number (or coefficient) of two disjoint curves $\gamma_i(t)\, i = 1, 2$ lying in the Euclidean space $\mathbb{R}^3$ $(\gamma_i(t) = r_i(t))$ $(0 \geq t \geq 2\pi$, r is the radius-vector of a point in $\mathbb{R}^3)$ is the number*

$$\{\gamma_1, \gamma_2\} = \frac{1}{4\pi} \int_{\gamma_1} \int_{\gamma_2} \frac{\langle [\mathrm{d}r_1, \mathrm{d}r_2], r_1 - r_2 \rangle}{|r_1 - r_2|^3}. \tag{10.5}$$

Here $\langle\ \rangle$ and $[\ ]$ denote scalar and vector product, respectively. It is possible to determine linking number coefficient in terms of homology. It will be followed from the equivalent definition the integer-valued of $\{\gamma_i, \gamma_j\}$ in (10.5).

Let us remind one important definition. We require several facts from the intersection theory, which we determine in more general setting.

10.1.3 Intersection Number

Let P^r and Q^s be two closed submanifolds of M^n of dimensions r and s, respectively. By the classical theorem of Whitney we can always regard M as an Euclidean space of a sufficiently large dimension.

P^r is said to intersect transversally to Q^s (or to be in general position) if at any point $x \in P^r \cap Q^s$, the tangent spaces $T_x P^r$ and $T_x Q^s$ generate the tangent space of M^n.

In particular, it follows that in general position the intersection $P^r \cap Q^3$ is a smooth $(r + s - n)$-dimensional submanifold.

Example 1. A straight line P^1 and the plane Q^2 in M^3 intersect transversally if P^1 does not meet Q^2 at the zero angle, i.e., is not in Q^2. If $\dim P^r + \dim Q^s = \dim M^n$ then in general position the manifolds P^r and Q^s at one or more points. If M^n, P^r, and Q^s are oriented, then each intersection point x_i is given a sign by the following rule: let τ_j^r be the oriented tangent frame to P^r at the point x_j and τ_j^s oriented frame to Q^s at x_j. The point x_j is assigned a plus sign if the union frame $(\tau_j^r, \tau_j^s$ is orientating for M^n at x_j. Otherwise, a minus sign is given. The sign is denoted by $\mathrm{sgn}_{x_j}(P \circ Q)$.

Definition 2. *The intersection number of two manifolds, P^r and Q^s, is the integer*

$$\mathrm{Ind}(P \circ Q) = \sum_{j=1}^{m} \mathrm{sgn}_{x_j}(P \circ Q), \tag{10.6}$$

where m is the number of intersection points.

In the nonorientable case Ind(P°Q) is defined as the residue module 2 of the number of intersection points.

Proposition 2. *Let f be a map of two-dimensional disk $D^2 \to \mathbb{R}^3$ that coincides with γ_1 on the boundary $\partial D^2 = S^1$ and is in general position on γ_2. Then* $\mathrm{Ind}(f(D^2) \cap \gamma_2)$ *is equal to* $\{\gamma_1, \gamma_2\}$.

Outline of the proof. The closed curves $\gamma_1(t)$ and $\gamma_2(t)$ define a two-dimensional oriented surface $\gamma_1 \times \gamma - 2 : (t_1, t_2) = (r_1(t_1), r_2(t_2))$ in $\mathbb{R}^6$. Let γ_1 and γ_2 be disjoint. Then the map (the so-called Gauss map)

$$\varphi(t_1, t_2) = \frac{r_1(t_1) - r_2(t_2)}{|r_1(t_1) - r_2(t_2)|}$$

is defined, with the degree given by integral (10.5). Therefore $\deg \varphi$ is an integer. $\deg \varphi$ remaining unaltered under nonintersecting deformation of curves γ_1 and γ_2, i.e., the linking coefficient $\{\gamma_1, \gamma_2\}$ is invariant under homotopies of γ_i. Since the intersection number in (10.6) depends linearly on points x_i it is suffices to calculate Ind (10.6) in two cases (a) γ_1 and γ_2 are unlinked and (b) γ_1 and γ_2 are two orthogonally linked circles (γ_1 is in the (x, y) plane and γ_2 in the (y, z) plane).

10.2 Hopf Invariant

10.2.1 Definition of Hopf Invariant

Let $f : S^3 \to S^2$ be a Hopf fibration. Consider two regular points a and b on the sphere S^2 and take their preimages $l_a^1 = f^{-1}(a)$ and $l_b^1 = f^{-1}(b)$. The manifolds l_a^1 and l_b^1 are two closed curves in S^3. Consider the linking number $\{l_a^1, l_b^1\}$.

Definition 3. *The Hopf invariant $h(f)$ is the linking number $\{l_a^1, l_b^1\}$.*

Proposition 3. *$h(f)$ is the homotopic invariant of f and is independent of the choice of points a and b.*

Proposition 3 is valid in a considerably more general situation of $2n - 1$-dimensional Hopf fibration $S^{2n-1} \to S^2$. The Hopf invariant $h(f)$ is defined for a $2n - 1$-dimensional Hopf fibration similar to the fiber bundle $S^3 \to S^2$. It should only be noticed that inverse of preimages of two points in $2n - 1$-dimensional sphere are $(n - 1)$-dimensional closed submanifolds M_1^{n-1} and M_2^{n-1}. Outline of proof of Proposition 3.1. We show that $h(f)$ is unaltered under a homotopy of f. Let f_0 and f_1 be mutually homotopic maps $S^{2n-1} \to S^n$ and $f - t$ the connecting homotopy. To prove that

$h(f_0) = h(f_1)$ it suffices to show that the deformation $f_t : S^{2n-1} \times I \to S^n$ connecting f_0 to f_1 and not passing through the points a and b can be constructed. Then the submanifolds $f_t^{-1}(a)$ and $f_t^{-1}(b)$ do not non-intersect under the homotopy; therefore the linking number is unaltered. 2. The independence of $h(f)$ from the choice of regular values a and b in S^n is proved quite simply. Let a_1 and b_1 two other points in S^n. There exists a map $\gamma : S^n \to S^n$ the sphere onto itself, homotopic to the identity (since $\pi_1(S^n) = 0$) and such $\gamma(a) = a_1$, $\gamma(b) = b_1$. Then the maps f and γf are homotopic. Therefore $h(f) = h(\gamma f)$.

10.2.2 Integral Representation of the Hopf Invariant

The Hopf invariant $h(f)$ admits a remarkable integral representation due to Whitehead [3]. This result has important applications in magneto-hydrodynamics, field theory, condensed matter. The Hopf invariant act as the topological conversation law.

First we formulate the Whitehead result for the classical Hopf fibration.

Proposition 4. *Let w^2 be a normalized 2-form on S^2, i.e., $\int_{S^2} w^2 = 1$ and $f : S^3 \to S^2$ is a smooth map. Consider the 2-form $\Omega^2 = f_*(w^2) = \mathrm{d}\xi^1$, where ξ^1 is a 1-form. Then*

$$\int_{S^3} f_*(w^2) \wedge \xi^1 = h(f). \tag{10.7}$$

The multidimensional generalization of the Whitehead formula for the Hopf fibration $S^{2n-1} \to S^n$ is the following:

Proposition 5. *Let ω^n be a normed n-form on S^n, i.e., $\int_{S^n} \omega^n = 1$. Then exists the n-form $f_*\omega^n$ on S^{2n-1} induced by f and exact, i.e., $f_*\omega^n = \mathrm{d}\xi^{n-1}$, where ξ^{n-1} is a $n-1$ - form on S^{2n-1}. Then*

$$\int f_*(\omega^n) \wedge \xi^{n-1} = h(f). \tag{10.8}$$

The proof of (10.7) and (10.8) can be found in [4]. (See also [2].)

10.3 Applications of Hopf Invariant

Two definitions of Hopf invariants via linking coefficients and integral formula (10.7) intimately linked among themselves and admit different generalizations and applications in many mathematical and physical problems. I discuss some examples that find or that could find some applications in biology. (See also the other chapters of this book.)

10.3.1 Generalized Linking Number

Consider the following problem: What kind of topological invariants would make it possible whether one can decouple (using motions in $\mathbb{R}^3$) a system of linked closed curves (loops). A classical invariant of this type is the Gauss-linking coefficient (10.5) of two loops. But knowing this coefficient is not enough to solve the problem of decoupling. Well-known examples such as the Whitehead link and Borromean rings (Fig. 10.2a, b) show that the condition $k_G(l_1, l_2) = 0$ gives only a necessary condition for decoupling the two curves. Such invariants are the high-order linking coefficients, which generalize the Gauss coefficients and as constructed in [5]. We show what the high-order linking coefficients look like in the simplest case of curves $l = (l_1, l_2, l_3)$ embedded in S^3. Let us begin with the coefficient $k(l_1, l_2)$ for two curves l_1 and l_2. As follows from section I the coefficient k can by defined by the two equivalent definitions:

1. As the intersection number $\mathrm{Ind}(z, l_2)$ of two-dimensional circle z (z is a film spanned on the curve l_1) with l_2 or
2. In the integral form (10.5)

The linking coefficient $k(l_1, l_2)$ defined in (10.5) can be calculated via differential forms u_1, u_2 defined on the curves l_1 and l_2. Forms u_i are defined by means of Alexander duality, which we determine in the special case one-dimensional sets embedded in S^3.

Proposition 6 (Alexander duality). *Let K be a one-dimensional compact set, $K \subset S^3$. There exists the isomorfphism f*

$$H_1(S^3 \backslash K) = H^1(K).$$

Let us apply the Alexander duality in the case $k = l = (l_1, l_2)$. The differential Alexander-dual 1-forms u_i is defined in the complement to l_i, closed and characterized by $\int_c u_i = k(c, l_i)$ for any closed curve from the complement of l_i in S^3. The cohomoly class of u_i is determined uniquely.

Now let B_i ($i = 1, 2$) be the boundary of tubular neighborhood of l_i not meeting another curve. Then

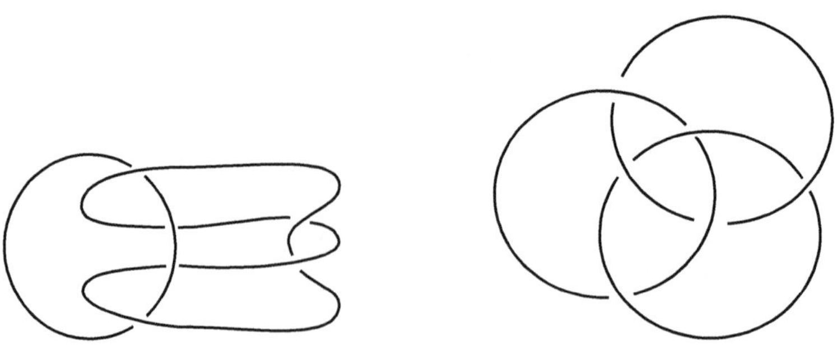

Fig. 10.2. (a) Whitehead link, (b) Borromean rings

$$\int_{B_1} u_1 \wedge u_2 = -\int_{B_2} u_2 \wedge u_1 = k(l_1, l_2). \tag{10.9}$$

Let us write the linking coefficient $k(l_1, l_2)$ using the first definition via intersection number. In fact it is contained in the Whitehead theorem (Proposition 4). We associate l_i with a closed 2-form v_i with support outside l_i. The form v_i is determined by the equality

$$\int_z v_i = \operatorname{Ind}(z, l_i).$$

Here z is an arbitrary 2-cycle in the complement to l_i.

3-forms $u_1 \wedge v_2$ and $v_1 \wedge u_2$ are defined on the whole sphere S^3 and

$$\int_{S^3} u_1 \wedge v_2 = \int_{S^3} k(l_1, l_2). \tag{10.10}$$

The proof of (10.10) is actually equivalent to the proof of (10.7).

The tuple of $k(l_i, l_j)$ is one of the numerical characteristics of the link $l = l_1, \dots, l_n$. It is also natural to introduce the linking number for the whole of l by the formula

$$\bar{k}(l) = \max_{1 \le i < j \le n} |k(l_i, l_j)|.$$

If l is isotopically unlinked, then $\bar{k}(l) = 0$. However for the links in Fig. 10.2a Whitehead link and 10.2b Borromean rings $\bar{k}(l) = 0$, but they remain linked. So we should introduce high-order linking numbers. If $\bar{k}(l) = 0$, where $l = (l_1, l_2)$, then there exists a 1-form u_{12} on the complement to l- and 2-forms v_{12}, v'_{12} with compact support on S^3 such that

$$du_{12} = u_1 \wedge u_2, \ dv_{12} = v_1 \wedge u_2, \ dv'_{12} = u_1 \wedge v_2.$$

Assume that $\bar{k}(l) = 0$ for $l = (l_1, l_2, l_3)$. In the addition to the above, we define 1-forms u_3 and u_{23} and 2-forms with compact support v_3, v_{23}, v'_{23} and verify by differentiation that the 2-form $\tilde{u}_{123} = u_{12} \wedge u_3 + u_1 \wedge u_{23}$ is closed. $\tilde{u}_{123}$ is defined on the complement of $l = (l_1, l_2, l_3)$. We also check by differentiation that the 3-forms

$$\tilde{v}_{123} = v_{12} \wedge u_3 + v_1 \wedge u_{23}, \ \tilde{u}'_{123} = u_{12} \wedge v_1 + u_1 \wedge v_{23}$$

are closed. v_{12} and v_{23} can be picked so that the latter 3-forms are defined on the whole sphere S^3.

The cohomology classes in $H^2(S^3 \backslash l)$ and $H^3(S^3)$ determined by $\tilde{u}_{123}$, $\tilde{v}_{123}$, and $\tilde{v}'_{123}$ are called the Massey products of cohomology classes $u_1, u_2, u_3, v_1, u_2, u_3, u_1, u_2, v_3$, denoted by $\langle clu_1, clu_2, clu_3 \rangle, \dots$, respectively. They do not depend on the choice of u_i, v_j, in clu_i, clv_i.

Proposition 7. *The integrals*

$$\int_{B_1} \tilde{u}_{123} = -\int_{B_2} \tilde{u}_{123} = \int_{S^3} \tilde{v}'_{123} = k_2(l) \tag{10.11}$$

have integer values and do not depend on the choice of $u_{12}, u_{23}, v_{12}, v_{23}$ *in the corresponding cohomology class.*

Let us consider sublinks of three curves $l_{ijk} = (l_i, l_j, l_k)$ for $1 \leq i < j \leq n$ of a link $l = (l_1, \ldots, l_n)$ with $n \geq 3$. We introduce the second linking number for l

$$\bar{k}_2(l) = \max_{1 \leq i < j < k \leq n} |k_2(l_i jk)|.$$

Example 1. For Borromean rings $k_2(l) = 1$. If l is homotopically unlinked, e.g., Whitehead link, then the number $\bar{k}_2(l)$ is zero. But there exist the links with $\bar{k}(l) = 0$, but homotopically unlinked. To characterize such links it is possible to define a linking number of order three ([2, 5]).

10.3.2 Formula Călugăreanu and Supercoiled DNA

Let γ be a closed smooth curve in $\mathbb{R}^3$ and v a normal vector field on γ. v is a one-parameter family of oriented line segments, one end of each which lies on γ, so that v forms a ribbon. We choose the length of γ to be so small that the line segment meets the curve γ only at its initial point, so that the ribbon is embedded. The curve of the endpoints of v γ_v inherits the orientation of the curve γ. Let $k(\gamma, \gamma_v)$ be the Gauss linking number of γ and γ_v. We define also the total twist of v

$$t_w = \frac{1}{2\pi} \int v^{\perp} \mathrm{d}v$$

in a standard way, where the vector $v^{\perp}$ is in the right-handed frame $(\hat{t}, v, v^{\perp})$, $\hat{t}$ is the unit tangent vector to the curve γ. The twist of the curve is a continuous quantity, the linking number is integer, therefore $k \neq t_w$. It is remarkable that another quantity can be introduced, viz., $k - t_w = Wr$ (the so-called writhing number) such that

$$k = t_w + Wr.$$

Wr only depends on γ and its definition is based on the following. We construct the Gauss map for $\gamma \times \gamma$, i.e., we define

$$\varphi : \gamma \times \gamma \to S^2, \quad \varphi(s, y) = \frac{y - x}{|y - x|}$$

for ordered pairs (x, y). Let $\mathrm{d}s$ be the element of area on S^2. Then the form $\varphi_*(\mathrm{d}s)$ is induced by φ.

Definition 4. *The writhing number is the integral*

$$Wr = \frac{1}{4\pi} \int_{\gamma} \int_{\gamma} \varphi_*(\mathrm{d}s). \tag{10.12}$$

Wr is a continuous quantity.

Proposition 8 (Formula Călugăreanu).

$$k(\gamma, \gamma_v) = t_w + Wr. \tag{10.13}$$

It is useful to compare the formula (10.5) for linking number and the formula (10.12) for writhing. The formal difference is that in the first case we integrate untegrand (10.5) over the space $\gamma \times \gamma_v$ and in the second one over the space $\gamma \times \gamma$. So the writhing number characterizes the single curve γ.

Remark 1. Wr is a very important quantity since in experiments with DNA molecule Wr can be measured directly, while k and t_w cannot be [22]. Different applications of formula (10.13) will be discussed in several chapters of our book. These applications and the modern proof of formula Călugăreanu can be found in the book [6]

Remark 2. Sometimes formula (10.13) is called formula Călugăreanu-White, or simply White formula. White proved in 1968 the multidimensional generalization of formula (10.13). In the article [23] he very clearly explained the relation between his result and Călugăreanu's [7]. But lately, in modern literature, the name of Călugăreanu gradually disappeared. (In this connection see the paper [8]).

Remark 3. The Călugăreanu formula implies that a ribbon swept by v forms a simple knot, i.e., the linking number $k(\gamma, \gamma_v)$ is the only topological invariant. It would be interesting to modify formula Călugăreanu in the case when the first linking number is zero and there exists next nontrivial second-order coefficient.

The question arises as to whether there are similar formulas for a linking ensemble $l = (l_1, l_2, \ldots, l_n)$.

10.3.3 Hopf Fibration and Membranes

One more example where Hopf fibration appears in physics and biology is the problem of phase transition in liquid membranes. We remind the mathematical essence of this problem. Applications in biological and physical systems can be found in [9, 10]. Let M^2 be a compact surface embedded in $\mathbb{R}^3$ and F the functional

$$F = \int_{M^2} H^2 \mathrm{d}A, \tag{10.14}$$

where H is the mean curvature.

Problem 1. To determine all compact surfaces of fixed genus g that minimize the functional (10.14)

$$\delta F = 0 \tag{10.15}$$

For arbitrary genus it is extremely difficult and remains an unsolved problem. There exist some partial results and interesting hypotheses. However for surfaces of genus $g = 0$ and $g = 1$ we have more complete results.

The following results due to Blaschke and Thomsen [11] are important.

Proposition 9. *Let $\tilde{M}^2$ be a minimal surface in S^3 and γ a stereographic projection $S^3 \to \mathbb{R}^3$. Then*

$$\gamma(\tilde{M}^2) = M^2 \textit{ and } F(M^2) = \gamma(\tilde{M}^2),$$

where $\delta(\tilde{M}^2)$ is the area of minimal surface.

Unfortunately, membranes are in general not exhausted by the projection of minimal surfaces. For example, there exists an infinite set of torical membranes that are not equivalent to minimal tori in S^3. These examples were constructed by Pinkal and used Hopf fibrations [12]. We called them *Pinkal Hopf tori.*

10.3.4 Construction of Hopf tori

We identify S^3 with the set of unit quaternions $\{a \in \mathbb{H},\, q\bar{q} = 1\}$ and S^2 with the unit sphere in the subspace of $\mathbb{H}$, spanned by $1, j, k$ but sends i to $-i$. Define

$$\pi : S^3 \to \mathbb{H} \quad \text{by } \pi(q) = \tilde{q}q.$$

Then π has the following properties:

(a) $\pi(S^3) = S^2$

(b) $\pi(\mathrm{e}^{\mathrm{i}\varphi}q) = \pi(q)$ for all $q \in S^3$, $\varphi \in \mathbb{R}$

(c) the group S^3 acts isometrically on S^3 by right multiplication and on S^2 via

$$q \mapsto \tilde{r}qr, \quad r \in S^3$$

π intertwines these two actions, i.e., for all $q, r \in S^3$ we have

$$\pi(qr) = \tilde{r}\pi(q)r.$$

Let $\rho : [a, b] \to S^2$ be an immersed curve. Choose $\eta : [a, b] \to S^3$ such that $\pi \circ \eta = \rho$. Let us consider an immersion χ of the cylinder $[a, b] \times S^1$ into S^3 by

$$\chi(t, \varphi) = \mathrm{e}^{\mathrm{i}\varphi}\eta(t). \tag{10.16}$$

If ρ is a closed curve, i.e., $\rho(t + L) = \rho(t)$, equation (10.16) determined a torus in S^3. This torus will be called the *Hopf torus* corresponding to ρ.

The main curvature $\tilde{H}$ of a Hopf torus coincides with the curvature k of the curve ρ.

The functional F is reduced to the integral

$$\pi \int_0^L (1 + k^2(s))\mathrm{d}s.$$

The article [13] has shown that there are infinitely many simple closed curves on S^2 that are critical points for F. Therefore there are many embedded Hopf tori in $\mathbb{R}^3$ that are not a minimal torius in S^3.

Problem 2. Is it possible to construct "critical" surface $\tilde{M}^2$ of genus g (solutions of (10.15)) gluing of Hopf tori.

Conclusion

An appearance of topological invariants like the Hopf invariant in such different problems of mathematics, physics, and even biology is far from being occasional. It reflects that the background of many modern constructions is based on the common topological ideas.

We have no opportunity to deeper in the relevant theories. We shall give only a short list of the topics with some references where a reader will be able to find both applications and generalizations of Hopf theory and a theme for the investigations:

1. The theory of multivalued functionals [14, 15] and the chapter by Millionschikov in this book.
2. Topological field theory [16, 24].
3. Fractional Statistics and quantum Hall effect [17, 18].
4. Topological invariants in magnetohydrodynamics [19, 20].
5. Knotlike configurations in relativistic field theory [21].
6. Quantum computations [25].

References

1. B. Dubrovin, S. Novikov, A. Fomenko. *Modern Geometry, I, II, III* (Springer, New York 1984–1990)
2. M. Monastyrsky, *Topology of Gauge Fields and Condensed Matter* (Plenum, New York, 1993)
3. J.H.C. Whitehead, Proc. Nat. Acad. Sci. USA, 117–123 (1947)
4. H. Whitney, *Geometric Integration theory* (Princeton University Press, Princeton, 1957)
5. M. Monastyrsky, V. Retakh, Commun. Math. Phys. **103**, 445–459, (1986)
6. J. White, Geometry and topology of DNA–protein interactions, in *New Scientific Applications of Geometry and Topology*, ed. by De Witt, L. Sumners, Proc. of Symposia in Applied Math. N 45 (AMS, Providence 1992)
7. G. Călugăreanu, Czechoslovak Math. J. **11**, 588–625 (1961)
8. H.H. Moffatt, R.L. Ricca, Proc. R. Soc. London A**439**, 411–429 (1992)
9. S.A. Safran *Statistical Thermodynamics of Surfaces, Interfaces, and Membrans* (Addison-Wesley, Reading, 2003)
10. E.I. Kats, M.I. Monastyrsky, JETP, **91**, 1279–1285 (2000)

11. W. Blaschke, G. Thomsen, *Vorlesungen über Differential geometry III* (Springer, Berlin Heidelberg New York, 1925)
12. U. Pinkal, Invent. Math. **81**, 379–386 (1985)
13. J. Langer, D. Singer, Bull. London Math. Soc. **16**, 531–534 (1984)
14. S.P. Novikov, Russ. Math. Survey, **37**, (N5), 349 (1982)
15. S.P. Novikov, Russ. Math. Survey, **39**, (N5), 17–106 (1984)
16. E. Witten, Commun. Math. Phys. **92**, 455–472 (1984)
17. F. Wilczek, *Fractional Statistics and Anyon Superconductivity* (World Scientific, Singapure, 1990)
18. R.G. Muf (J. Frölich et al.) The Fractional Quantum Hall effect, Chern-Simons theory, and Integral lattice, Proceedings of International Congress of Mathematicians, (Zürich, Birkhäuser, Basel, 1995)
19. M.I. Monastyrsky, P.V. Sasorov, Sov. Phys. JETP **66**(4), 683–688 (1987)
20. V.I. Arnold, B.A. Khesin. *Topological Methods in Hydrodynamics*, (Springer, Berlin Heidelberg New York, 1998)
21. L. Faddeev, A.J. Niemi, Toroidal configurations as stable solitons, hep-th 9705176 (22.05.1997)
22. F.B. Fuller, Proc. Nat. Acad. Sci. USA **75**, 3557 (1978)
23. J. White, Amer. J. Math. **91**, 693–728 (1969)
24. M.F. Atiyah, The geometry and physics of knots (Cambridge University Press, Cambridge, 1990)
25. R. Mosseri, Two and three qubits geometry and Hopf fibrations, (in Topology and Condensed matter, ed. by M. Monastyrsky (Springer, Berlin Heidelberg New York 2005)

11

Multi-Valued Functionals, One-Forms and Deformed de Rham Complex*

D.V. Millionschikov

Summary. We discuss some applications of the Morse–Novikov theory to some problems in modern physics, which appears as a non-exact closed 1-form ω (multi-valued functional). We focus our attention mainly on the cohomology $H^*_{\lambda\omega}(M^n, \mathbb{R})$ of the de Rham complex $\Lambda^*(M^n)$ of a compact manifold M^n with a deformed differential $\mathrm{d}_\omega = \mathrm{d} + \lambda\omega$. Using Witten's approach to the Morse theory one can estimate the number of critical points of ω in terms of $H^*_{\lambda\omega}(M^n, \mathbb{R})$ with sufficiently large values of λ (torsion-free Novikov's inequalities).

We show that for an interesting class of solvmanifolds G/Γ the cohomology $H^*_{\lambda\omega}(G/\Gamma, \mathbb{R})$ can be computed as the cohomology $H^*_{\lambda\omega}(\mathfrak{g})$ of the corresponding Lie algebra $\mathfrak{g}$ associated with the one-dimensional representation $\rho_{\lambda\omega}$. Moreover $H^*_{\lambda\omega}(G/\Gamma, \mathbb{R})$ is almost always trivial except a finite number of classes $[\lambda\omega]$ in $H^1(G/\Gamma, \mathbb{R})$.

11.1 Introduction

In the beginning of the 1980s Novikov constructed [1, 2] an analogue of the Morse theory for smooth multi-valued functions, i.e. smooth closed 1-forms on a compact smooth manifold M. In particular he introduced the Morse-type inequalities (Novikov's inequalities) for numbers $m_p(\omega)$ of zeros of index p of a Morse 1-form ω.

In [3, 4] a method of obtaining the torsion-free Novikov inequalities in terms of the de Rham complex of manifold was proposed. This method was based on Witten's approach [5] to the Morse theory. Pazhitnov generalized Witten's deformation $\mathrm{d} + t\mathrm{d}f$ (f is a Morse function on M) of the standard differential d in $\Lambda^*(M)$ by replacing $\mathrm{d}f$ by an arbitrary Morse 1-form on M. For sufficiently large real values t one has the following estimate (torsion-free Novikov inequality [4, 6]):

$$m_p(\omega) \geq \dim H^p_{t\omega}(M, \mathbb{R}),$$

* Partially supported by the Russian Foundation for Fundamental Research, grant no. 99-01-00090 and PAI-RUSSIE, dossier no. 04495UL

where by $H^p_{t\omega}(M,\mathbb{R})$ we denote the p-th cohomology group of the de Rham complex $(\Lambda^*(M), \mathrm{d}+t\omega)$ with respect to the new deformed differential $\mathrm{d}+t\omega$.

Taking a complex parameter λ one can identify $H^*_{\lambda\omega}(M,\mathbb{C})$ with the cohomology $H^*_{\rho_{\lambda\omega}}(M_{\mathrm{n}},\mathbb{C})$ with coefficients in the local system $\rho_{\lambda\omega}$ of groups $\mathbb{C}$, where

$$\rho_{\lambda\omega}(\gamma) = \exp\int_\gamma \lambda\omega, \gamma \in \pi_1(M).$$

Alania in [7] studied $H^*_{\rho_{\lambda\omega}}(M_{\mathrm{n}},\mathbb{C})$ of a class of nilmanifolds M_{n}. He proved that $H^*_{\rho_{\lambda\omega}}(M_{\mathrm{n}},\mathbb{C})$ is trivial if $\lambda\omega \neq 0$. The proof was based on the Nomizu theorem [8] that reduces the problem in the computation in terms of the corresponding nilpotent Lie algebra. It was remarked in [9, 10] that triviality of $H^*_{\rho_{\lambda\omega}}(G/\Gamma,\mathbb{R})$, with $\lambda\omega \neq 0$ follows from Dixmier's theorem [11], namely: *for a nilmanifold G/Γ the cohomology $H^*_\omega(G/\Gamma,\mathbb{R})$ coincides with the cohomology $H^*_\omega(\mathfrak{g})$ associated with the one-dimensional representation of the Lie algebra $\rho_\omega : \mathfrak{g} \to \mathbb{R}, \rho_\omega(\xi) = \omega(\xi)$ and hence $H^*_\omega(G/\Gamma,\mathbb{R}) = H^*_\omega(\mathfrak{g}) = 0$.*

Applying Hattori's theorem [12] one can observe that the isomorphism

$$H^*_\omega(G/\Gamma,\mathbb{R}) \cong H^*_\omega(\mathfrak{g})$$

still holds on for compact solvmanifolds G/Γ with completely solvable Lie group G. The calculations show that the cohomology $H^*_\omega(G/\Gamma,\mathbb{R})$ can be non-trivial for certain values $[\omega] \in H^1(G/\Gamma,\mathbb{R})$. However there exist only a finite number of such values.

Let us consider a finite subset $\Omega_{G/\Gamma}$ in $H^1(G/\Gamma,\mathbb{R}) \cong H^1(\mathfrak{g})$:

$$\Omega_{G/\Gamma} = \{\alpha_{i_1}+\cdots+\alpha_{i_s} |\ 1 \le i_1 < \cdots < i_s \le n,\ \ s = 1,\ldots,n\},$$

where the set $\{\alpha_1,\ldots,\alpha_n\}$ of closed 1-forms is in fact the set of the weights of completely reducible representation associated to the adjoint representation of $\mathfrak{g}$. It was proved in [9]: *if $-[\omega] \notin \Omega_{G/\Gamma}$, then the cohomology $H^*_\omega(G/\Gamma,\mathbb{R})$ is trivial.*

11.2 Dirac Monopole, Multi-Valued Actions and Feynman Quantum Amplitude

The notion of multi-valued functional originates from topological study of the quantization process of the motion of a charged particle in the field of a Dirac monopole [13]. The Kirchhoff–Thomson equations for free motion of solids in a perfect noncompressible liquid also can be reduced to the theory of a charged particle on the sphere S^2 with some metric $g_{\alpha\beta}$ in a potential field U and in an effective magnetic field $F = F_{12}$ with a non-zero flux $4\pi s$ through S^2. Locally (in some domain U_α) on the sphere we have the following formula for the action $S_\alpha(\gamma)$:

$$S_\alpha(\gamma) = \int_\gamma \left(\frac{1}{2} g_{ij}\dot{x}^i\dot{x}^j - U + eA_k^\alpha \dot{x}^k\right) \mathrm{d}t, \tag{11.1}$$

where

$$x^1 = \theta,\ x^2 = \varphi, \quad F_{12}\mathrm{d}\theta \wedge \mathrm{d}\varphi = \mathrm{d}(A_k^\alpha \mathrm{d}x^k), \quad \int\int_{S^2} F_{12}\mathrm{d}\theta \wedge \mathrm{d}\varphi = 4\pi s \neq 0. \tag{11.2}$$

One can consider Feynman's paths integral approach to the quantization of the problem considered above. Recall that in the standard situation of single-valued action S, we consider the amplitude

$$\exp\{2\pi \mathrm{i} S(\gamma)\}, \quad \gamma \in \Omega(x, x')$$

and the propagator

$$K(x, x') = \int_{\Omega(x,x')} \exp\{2\pi \mathrm{i} S(\gamma)\} \mathrm{D}\gamma.$$

For the Dirac monopole one can consider the set $\{S_1, S_2\}$ of local actions where $U_1 = S^2 \backslash P_\mathrm{N}$ and $U_2 = S^2 \backslash P_\mathrm{S}$, by $P_\mathrm{N}, P_\mathrm{S}$ we denote the poles of the sphere S^2. Taking the equator γ with the positive orientation, one can easily test the ambiguity of the action:

$$S_1(\gamma) - S_2(\gamma) = e\int_\gamma (A_k^1 \mathrm{d}x^k - A_k^2 \mathrm{d}x^k) = e\int\int_{S^2} F_{12}\mathrm{d}\theta \wedge \mathrm{d}\varphi = 4\pi s e \neq 0. \tag{11.3}$$

The monopole is quantized if and only if the amplitude $\exp\{2\pi \mathrm{i} S_\alpha(\gamma)\}$ is a single-valued functional, i.e. for an arbitrary closed $\gamma \in U_1 \cap U_2$ we have

$$\exp\{2\pi \mathrm{i} S_1(\gamma)\} = \exp\{2\pi \mathrm{i} S_2(\gamma)\}.$$

The last condition is equivalent to the following one:

$$4\pi s e = k, \ k \in \mathbb{Z}. \tag{11.4}$$

Generalizing the situation with the Dirac monopole Novikov [2] considered an n-dimensional manifold $M^n, n > 1$ with a metric g_{ij}, with a scalar potential U and with a two-form F of magnetic field not necessarily exact. In these settings one can consider a set of open $U_\alpha \subset M^n$, such that $F = F_{ij}\mathrm{d}x^i \wedge \mathrm{d}x^j$ is exact on U_α and $M^n \subset \cup_\alpha U_\alpha$. A 1-form $\omega_\alpha = A_k^\alpha \mathrm{d}x^k$, $\mathrm{d}\omega_\alpha = F_{ij}\mathrm{d}x^i \wedge \mathrm{d}x^j$ is determined up to some closed 1-form and we can consider the set of local actions:

$$S_\alpha(\gamma) = \int_\gamma \left(\frac{1}{2} g_{ij}\dot{x}^i\dot{x}^j - U + eA_k^\alpha \dot{x}^k\right) \mathrm{d}t, \tag{11.5}$$

Let us consider a path $\gamma \subset U_\alpha \cap U_\beta$. The values $S_\alpha(\gamma)$ and $S_\beta(\gamma)$ do not coincide generally speaking. Hence the set $\{S_\alpha\}$ of local actions defines a multi-valued functional S. As $\omega_\alpha - \omega_\beta$ is closed on $U_\alpha \cap U_\beta$ the integral

$$S_\alpha(\gamma_\lambda) - S_\beta(\gamma_\lambda) = \int_{\gamma_\lambda} (\omega_\alpha - \omega_\beta)$$

is invariant under any deformation $\gamma_\lambda \subset U_\alpha \cap U_\beta$ of γ in the class: (a) periodic curves; (b) the curves with the same end points.

The crucial Novikov's observation was: the infinite-dimensional 1-form δS is well defined and closed for the following function spaces:

(a) Ω^+ of the oriented closed contours γ, such that $\exists \alpha, \gamma \subset U_\alpha$;

(b) $\Omega(x, x')$ of the paths $\gamma(x, x')$ joining points x, x', such that $\exists \alpha, \gamma(x, x') \subset U_\alpha$.

The set $\{S_\alpha\}$ of local actions determines also a multi-valued (in general) functional $\exp\{2\pi \mathrm{i} S\}$ on $\Omega(x, x')$. The local variational system $\{S_\alpha\}$ is quantized if and only if the Feynman quantum amplitude $\exp\{2\pi \mathrm{i} S\}$ is a single-valued functional on Ω^+. Or, in other words, for all $\gamma \in U_\alpha \cap U_\beta$ we have

$$\int_\gamma (\omega_\alpha - \omega_\beta) = k, \quad k \in \mathbb{Z}. \tag{11.6}$$

If U_α and U_β are simply connected domains in M^n it is possible to consider a map $f : S^2 \to M$ such that γ is the image of the equator of the sphere S^2 and the images of two half-spheres of S^2 lie in U_α and U_β, respectively. Then condition (11.6) can be rewritten as

$$\int_{f(S^2)} F_{ij} \mathrm{d}x^i \wedge \mathrm{d}x^j = k, \quad k \in \mathbb{Z}. \tag{11.7}$$

Hence we can propose the following sufficient condition of the quantization: *a local variational system $\{S_\alpha\}$ on M^n that corresponds to some magnetic field $F = F_{ij}\mathrm{d}x^i \wedge \mathrm{d}x^j$ is quantized if F has integer-valued fluxes through all basic cycles of $H_2(M^n, \mathbb{Z})$.*

One can remark that the last condition is in fact excessive: it is sufficient to require integer-valued integrals of F over spheric cycles that lie in the image of the Hurewicz map $H : \pi_2(M^n) \to H_2(M^n, \mathbb{Z})$.

11.3 Aharonov–Bohm Field and Equivalent Quantum Systems

Another interesting example comes from Aharonov–Bohm experiment [14–17]. We consider the electron moves outside the ideal endless solenoid, i.e. the configuration space is $M = (\mathbb{R}^2 \backslash D_\varepsilon) \times \mathbb{R} = \{(x, y, z) \in \mathbb{R}^3, x^2 + y^2 > \varepsilon^2\}$, where

D_ε is two-dimensional disk of radius $\varepsilon \to 0$. The magnetic field $F = F_{ij}\mathrm{d}x^i \wedge \mathrm{d}x^j$ vanishes outside solenoid, i.e. $F \equiv 0$ on M, hence

$$S_{\omega_\alpha}(\gamma) = \int_\gamma \frac{m\dot{x}^2}{2}\mathrm{d}t + \omega_\alpha, \tag{11.8}$$

where $\omega_\alpha = eA_k\mathrm{d}x^k$ is an arbitrary closed 1-form on M. The cohomology space

$$H^1(M,\mathbb{R}) = H^1(\mathbb{R}^2\backslash D_\varepsilon,\mathbb{R}) = H^1(S^1,\mathbb{R}) = \mathbb{R}$$

is one dimensional and hence

$$\omega_\alpha = \frac{e\Phi_\alpha}{2\pi}\frac{x\mathrm{d}y - y\mathrm{d}x}{x^2+y^2} + \mathrm{d}f_\alpha,$$

for some constant Φ_α and function f_α on M.

Taking the circle $\gamma_0 = \partial D_\varepsilon = \{(\varepsilon\cos\varphi, \varepsilon\sin\varphi, 0), 0 \le \varphi < 2\pi\}$ we have

$$\int_{\gamma_0} A_k^\alpha \mathrm{d}x^k = \frac{1}{e}\int_{\gamma_0}\omega_\alpha = \Phi_\alpha = \int_{D_\varepsilon} F_{12}\mathrm{d}x \wedge \mathrm{d}y.$$

Hence the constant Φ_α is equal to the flux of the magnetic field F through the orthogonal section D_ε of our solenoid.

The form ω_α determines a representation ρ_{ω_α} of the fundamental group of M:

$$\rho_{\omega_\alpha} : \pi_1(M) \to \mathbb{C}^*, \quad \rho_{\omega_\alpha}(\gamma) = \exp\left\{2\pi\mathrm{i}\int_\gamma \omega_\alpha\right\}, \quad \gamma \in \pi_1(M).$$

Let S_{ω_1} and S_{ω_2} be two actions for our system. They are quantum-mechanically equivalent if and only if

$$\exp\{2\pi\mathrm{i}S_{\omega_1}(\gamma)\} = c(x,x')\exp\{2\pi\mathrm{i}S_{\omega_2}(\gamma)\},$$

with a phase factor $c(x,x')$ depending only on end points x, x' of γ and $|c(x,x')| = 1$, i.e. $c(x,x')$ is physically unobservable. It is easy to show that the actions S_{ω_1} and S_{ω_2} are quantum-mechanically equivalent if and only if for any loop $\gamma \in \pi_1(M)$ the value of the integral $\int_\gamma(\omega_1 - \omega_2)$ is integer or, in other words, the form $(\omega_1 - \omega_2)$ has integer-valued integrals over basic cycles of $H_1(M,\mathbb{Z})$.

In our case $H_1(M,\mathbb{Z}) = \mathbb{Z}$ and the last condition is equivalent to the following one

$$\int_{\gamma_0}(\omega_1 - \omega_2) = e(\Phi_1 - \Phi_2) = k, \quad k \in \mathbb{Z}. \tag{11.9}$$

Hence (one of the important observations in the Aharonov–Bohm experiment) the fields with fluxes Φ_1 and Φ_2, such that $\Phi_1 - \Phi_2 = k/e, k \in \mathbb{Z}$ cannot be distinguished by any interference effect.

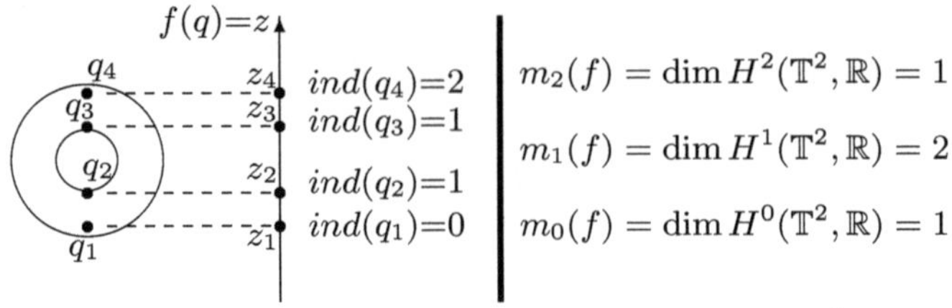

Fig. 11.1. A height-function $f(q) = z$ for $\mathbb{T}^2$

Now let us consider the case when M^n is not simply connected and the two-form F is globally exact on M^n (like in the Aharonov–Bohm experiment). Two solutions ω_1, ω_2 of the equation $\mathrm{d}\omega = F_{ij}\mathrm{d}x^i \wedge \mathrm{d}x^j$ that correspond to two different actions $S_1(\gamma)$ and $S_2(\gamma)$ are determined up to a differential $\mathrm{d}f$ by their integrals $\int_{\gamma_k} \omega_i$ over the basic cycles γ_k of $H_1(M^n, \mathbb{Z})$. These integrals can be interpreted as the fluxes of the continuation of F (with possible singularities) to some large manifold $\tilde{M}^n$. Two variational systems $S_1(\gamma)$ and $S_2(\gamma)$ are quantum-mechanically equivalent if and only if all integrals $\int_{\gamma_k}(\omega_1 - \omega_2)$ over basic cycles γ_k of $H_1(M^n, \mathbb{Z})$ are integer valued.

The form $\omega_{12} = \omega_1 - \omega_2$ is a closed 1-form on M^n and it determines a representation $\rho_{\omega_{12}}$ of the fundamental group $\pi_1(M^n)$:

$$\rho_{\omega_{12}} : \pi_1(M^n) \to \mathbb{C}^*, \quad \rho_{\omega_{12}}(\gamma) = \exp\left\{2\pi\mathrm{i}\int_\gamma \omega_{12}\right\}, \quad \gamma \in \pi_1(M^n).$$

Let M be a finite-dimensional (or infinite-dimensional) manifold and $S : M \to \mathbb{R}$ a function (functional) on it.

What are the relations between the set of the stationary points $\mathrm{d}S = 0$ ($\delta S = 0$) and the topology of the manifold M?

If S is a Morse function (generic situation), i.e. $\mathrm{d}^2 S$ is non-degenerate at critical points, then one can define the Morse index ind(P) of a critical point P of S as the number of negative squares of the quadratic form $\mathrm{d}^2 S(P)$ (if it is finite in the infinite-dimensional case) (Fig. 11.1) [22].

Under some natural hypotheses the following inequality can be established:

$$m_p(S) \geq b_p(M) = \dim H^p(M).$$

11.4 Semi-Classical Motion of Electron and Critical Points of 1-Form

The semi-classical model of electron motion is an important tool for investigating conductivity in crystals under the action of a magnetic field [18, 19]. At the same time it is one of the most important examples of applications of topological methods in the modern physics.

Let us consider the corresponding quantum system defined for some crystal lattice $L = \mathbb{Z}^3$. Its eigenstates are the Bloch functions ψ_p. The particle quasi-momentum p is defined up to a vector of the dual lattice $L^* = \mathbb{Z}^3$. Hence one can regard the space of quasi-momenta as a 3-dimensional torus $\mathbb{T}^3 = \mathbb{R}^3/\mathbb{Z}^3$. The state energy $\varepsilon(p)$ is thus a function on $\mathbb{T}^3$, i.e. a 3-periodical function in $\mathbb{R}^3$.

An external homogeneous constant magnetic field is a constant vector $H = (H_1, H_2, H_3)$ or in other words it is a 1-form $\omega = H_1 \mathrm{d}p_1 + H_2 \mathrm{d}p_2 + H_3 \mathrm{d}p_3$ with constant coefficients.

The semi-classical trajectories projected to the space of quasi-momenta are connected components of the intersection of the planes $(p, H) = \text{const.}$ with constant energy surfaces $\varepsilon(p) = \text{const.}$

The constant energy surfaces $\varepsilon(p) = \varepsilon_F$ that correspond to the Fermi energies ε_F are called *the Fermi surfaces.* There are non-closed trajectories on the Fermi surfaces with asymptotic directions and this topological fact explains an essential anisotropy of the metal conductivity at low temperatures.

One can study the intersections

$$(p, H) = c_0, \quad \varepsilon(p) = \varepsilon_0$$

as the level surfaces of the 1-form

$$\omega_{\varepsilon_0} = (H_1 \mathrm{d}p_1 + H_2 \mathrm{d}p_2 + H_3 \mathrm{d}p_3)|_{\hat{M}_{\varepsilon_0}},$$

where 2-dimensional manifold

$$\hat{M}_{\varepsilon_0} = \{p \in \mathbb{R}^3 | \varepsilon(p) = \varepsilon_0\}$$

is the universal covering of the compact Fermi surface $\varepsilon(p) = \varepsilon_0$ in $\mathbb{T}^3$. The last one is denoted also by M_{ε_0}. We can treat the 3-periodic form ω_{ε_0} as a 1-form on the compact manifold M_{ε_0} (we keep the same notation for it).

The information about critical points of ω_{ε_0} is very important in the problem considered earlier. A generic 1-form ω_{ε_0} is a Morse form and has finitely many critical points on M_{ε_0}.

11.5 Witten's Deformation of de Rham Complex and Morse–Novikov Theory

In 1982 Witten proposed a new beautiful proof of the Morse inequalities using some analogies with supersymmetry quantum mechanics [5]. Taking an arbitrary smooth real-valued function f on a Riemannian manifold M^n he considered a new deformed differential d_t in the de Rham complex $\Lambda^*(M^n)$ (t is a real parameter):

$$\begin{gathered} \mathrm{d}_t = \mathrm{e}^{-ft} \mathrm{d} \mathrm{e}^{ft} = \mathrm{d} + t \mathrm{d}f \wedge, \\ \mathrm{d}_t(\xi) = \mathrm{d}\xi + t\mathrm{d}f \wedge \xi, \ \xi \in \Lambda^*(M^n), \end{gathered} \tag{11.10}$$

where d is the standard differential in $\Lambda^*(M^n)$:

$$\begin{aligned} \mathrm{d} : \Lambda^p(M^n) &\to \Lambda^{p+1}(M^n), \\ \xi = \sum_{i_1<\dots<i_p} \xi_{i_1\dots i_p} \mathrm{d}x^{i_1} &\wedge \dots \wedge \mathrm{d}x^{i_p} \in \Lambda^p(M^n), \\ \mathrm{d}\xi = \sum_{i_1<\dots<i_p} \sum_q \frac{\partial \xi_{i_1\dots i_p}}{\partial x^q} \mathrm{d}x^q \wedge \mathrm{d}x^{i_1} &\wedge \dots \wedge \mathrm{d}x^{i_p} \in \Lambda^{p+1}(M^n). \end{aligned} \tag{11.11}$$

Taking arbitrary smooth vector fields $X_1, \dots, X_{p+1}$ on M^n we have also the following formula:

$$\begin{aligned} \mathrm{d}\xi(X_1, \dots, X_{p+1}) = \sum_{1 \le i<j \le p+1} (-1)^{i+j} \xi([X_i, X_j], X_1, \dots, \hat{X}_i, \dots, \hat{X}_j, \dots, X_{q+1}) \\ + \sum_i (-1)^{i+1} X_i \xi(X_1, \dots, \hat{X}_i, \dots, X_{p+1}). \end{aligned} \tag{11.12}$$

We recall that a differential p-form ξ is called *closed* if $\mathrm{d}\xi = 0$ and it is called *exact* if $\xi = \mathrm{d}\xi'$ for some $(p-1)$-form ξ'. As $\mathrm{d}^2 = 0$ the space of exact forms is a subspace of the space of closed ones and the p-th de Rham cohomology group $H^p(M^n, \mathbb{R})$ of the manifold M^n is defined as a quotient space of closed p-forms modulo exact ones. In the same manner the cohomology $H_t^*(M^n, \mathbb{R})$ of the de Rham complex with respect to the deformed differential d_t can be defined.

The operators d_t and d are conjugated by the invertible operator e^{ft} and therefore the cohomology groups $H^*(M^n, \mathbb{R})$ (the standard de Rham cohomology) and $H_t^*(M^n, \mathbb{R})$ (the new ones) are isomorphic to each other. On the level of the forms this isomorphism is given by the gauge transformation

$$\xi \to \mathrm{e}^{ft} \xi.$$

One can define the adjoint operator $\mathrm{d}_t^* = \mathrm{e}^{ft} \mathrm{d}^* \mathrm{e}^{-ft}$ with respect to the scalar product of differential forms

$$(\alpha, \beta) = \int_{M^n} (\alpha, \beta)_x \mathrm{d}V,$$

where $(\alpha, \beta)_x$ is a scalar product in the bundle $\Lambda^*(T_x^*(M^n))$ evaluated with respect to the Riemannian metric g_{ij} of M^n and $\mathrm{d}V$ is the corresponding volume form.

One can also consider the deformed Laplacian $H_t = \mathrm{d}_t \mathrm{d}_t^* + \mathrm{d}_t^* \mathrm{d}_t$ acting on forms. An arbitrary element ω from $H_t^p(M^n, \mathbb{R})$ can be uniquely represented as an eigenvector with zero eigenvalue of the Hamiltonian $H_t = \mathrm{d}_t \mathrm{d}_t^* + \mathrm{d}_t^* \mathrm{d}_t$. Hence one can compute the Betti number $b_p(M^n) = \dim H^p(M^n, \mathbb{R})$ as the number of zero eigenvalues of H_t acting on p-forms.

It can be calculated that

$$H_t = \mathrm{d}_t\mathrm{d}_t^* + \mathrm{d}_t^*\mathrm{d}_t = \mathrm{d}\mathrm{d}^* + \mathrm{d}^*\mathrm{d} + t^2(\mathrm{d}f)^2 + t\sum_{i,j}\nabla^2_{(i,j)}(f)[\tilde{a}^i, \tilde{a}^{j*}], \quad (11.13)$$

where $(\mathrm{d}f)^2 = (\mathrm{d}f, \mathrm{d}f)_x = g^{ij}\frac{\partial f}{\partial x^i}\frac{\partial f}{\partial x^j}$ and

$$\tilde{a}^i(\xi) = \mathrm{d}x^i \wedge \xi, \quad \nabla^2_{(i,j)} = \nabla_i\nabla_j - \Gamma^k_{ij}\nabla_k.$$

As the "potential energy" $t^2(\mathrm{d}f)^2$ of the Hamiltonian H_t becomes very large for $t \to +\infty$ the eigenfunctions of H_t are concentrated near the critical points $\mathrm{d}f = 0$ and the low-lying eigenvalues of H_t can be calculated by expanding about the critical points. Taking the Morse coordinates x^i in some neighbourhood W of a critical point P

$$f(x) = \frac{1}{2}\sum \lambda_i(x^i)^2, \quad \lambda_1 = \cdots = \lambda_q = -1, \quad \lambda_{q+1} = \cdots = \lambda_n = 1,$$

where q is the index of the critical point P, and introducing a Riemmanian metric g_{ij} on M^n such that x^i are Euclidean coordinates for g_{ij} in W one can locally evaluate the Hamiltonian H_t:

$$H_t = \sum_i \left(-\frac{\partial^2}{\partial x^{i2}} + t^2 x^{i2} + t\lambda_i[\tilde{a}^i, \tilde{a}^{i*}]\right). \quad (11.14)$$

The operator

$$H_i = -\frac{\partial^2}{\partial x^{i2}} + t^2 x^{i2}$$

is the Hamiltonian of the simple harmonic oscillator and it has the following set of eigenvalues

$$t(1 + 2N_i), \quad N_i = 0, 1, 2, \ldots$$

with simple multiplicities. The operator H_i commutes with $[\tilde{a}^i, \tilde{a}^{i*}]$ and the eigenvalues of the last operator are equal to ± 1:

$$[\tilde{a}^i, \tilde{a}^{i*}](\psi(x)\mathrm{d}x^{i_1} \wedge \ldots \wedge \mathrm{d}x^{i_p}) = \begin{cases} \psi(x)\mathrm{d}x^{i_1} \wedge \cdots \wedge \mathrm{d}x^{i_p}, & i \in (i_1, \ldots, i_p), \\ -\psi(x)\mathrm{d}x^{i_1} \wedge \cdots \wedge \mathrm{d}x^{i_p}, & i \notin (i_1, \ldots, i_p). \end{cases}$$

Hence the eigenvalues of the restriction $H_t|_W$ are equal to

$$t\sum_i(1 + 2N_i + \lambda_i l_i), \quad N_i = 0, 1, 2, \ldots, \; l_i = \pm 1. \quad (11.15)$$

The corresponding eigenfunctions $\Psi_t = \psi(x, t)\mathrm{d}x^{i_1} \wedge \cdots \wedge \mathrm{d}x^{i_p}$ are defined in W and not on the whole manifold M^n. Using the partition of unit one can define a new smooth q-form $\tilde{\Psi}_t$ on M^n such that $\tilde{\Psi}_t$ coincides with Ψ_t in some $\tilde{W} \subset W$ and $\tilde{\Psi}_t \equiv 0$ outside of W. The q-form $\tilde{\Psi}_t$ is called a quasi-mode:

$$H_t\tilde{\Psi}_t = t\left(\sum_i (1 + 2N_i + \lambda_i l_i) + \frac{B}{t} + \frac{C}{t^2} + \cdots\right)\tilde{\Psi}_t, \quad t \to +\infty. \tag{11.16}$$

The numbers $t\sum_i (1 + 2N_i + \lambda_i l_i)$ are called *asymptotic eigenvalues* and their minimal value E_0^{as} approximates the minimal eigenvalue of H_t as $t \to +\infty$.

In order to find E_0^{as}, we must set $N_i = 0$ for all i. The sum

$$\sum_{i=1}^{q}(1 - l_i) + \sum_{i=q+1}^{n}(1 + l_i).$$

is non-negative and it is equal to zero if and only if

$$l_1 = \cdots = l_q = 1, \quad l_{q+1} = \cdots = l_n = -1.$$

This means that H_t has precisely one zero asymptotic eigenvalue for each critical point of index q. Hence we have precisely $m_q(f)$ asymptotic zero eigenvalues (for q-forms). Vanishing of the first term of the asymptotical expansion (11.16) for a minimal eigenvalue of H_t is only a necessary condition to have zero energy level; hence the number $b_q(M^n)$ of zero eigenvalues does not exceed the number of zero asymptotic eigenvalues. In other words we have established the Morse inequalities

$$m_q(f) \geq b_q(M^n).$$

It was Pajitnov who remarked that it is possible to apply Witten's approach to the Morse–Novikov theory [4]. Let ω be a closed 1-form on M^n and t a real parameter. As in the construction earlier one can define a new deformed differential $\mathrm{d}_{t\omega}$ in $\Lambda^*(M)$

$$\mathrm{d}_{t\omega} = \mathrm{d} + t\omega\wedge, \quad \mathrm{d}_{t\omega}(\xi) = \mathrm{d}a + t\omega \wedge \xi.$$

If the 1-form ω is not exact, the cohomology $H^*_{t\omega}(M,\mathbb{R})$ of the de Rham complex with the deformed differential $\mathrm{d}_{t\omega}$ generally speaking is not isomorphic to the standard one $H^*(M,\mathbb{R})$. But $H^*_{t\omega}(M,\mathbb{R})$ depends only on the cohomology class of ω: for any pair ω, ω' of 1-forms such that $\omega - \omega' = \mathrm{d}\phi$, where ϕ is a smooth function on M^n; the cohomology $H^*_{t\omega}(M,\mathbb{R})$ and $H^*_{t\omega'}(M,\mathbb{R})$ is isomorphic to each other. This isomorphism can be given by the gauge transformation

$$\xi \to \mathrm{e}^{t\phi}\xi; \quad \mathrm{d} \to \mathrm{e}^{t\phi}\mathrm{d}\mathrm{e}^{-t\phi} = \mathrm{d} + t\mathrm{d}\phi \wedge .$$

It is convenient also to consider a complex parameter λ instead of t. It was remarked in [3, 4] that the cohomology $H^*_{\lambda\omega}(M,\mathbb{C})$ of $\Lambda^*(M)$ with respect to the deformed differential $d_{\lambda\omega}$ coincides with the cohomology $H^*_{\rho_{\lambda\omega}}(M,\mathbb{C})$

with coefficients in the representation $\rho_{\lambda\omega} : \pi_1(M) \to \mathbb{C}^*$ of fundamental group defined by the formula

$$\rho_{\lambda\omega}([\gamma]) = \exp \int_\gamma \lambda\omega, \quad [\gamma] \in \pi_1(M),$$

We denote corresponding Betti numbers by $b_p(\lambda, \omega)$, $b_p(\lambda, \omega) = \dim H^*_{\rho_{\lambda\omega}}(M, \mathbb{C})$.

There is another interpretation of $H^*_{\rho_{\lambda\omega}}(M, \mathbb{C}$: the representation $\rho_{\lambda\omega} : \pi_1(M) \to \mathbb{C}^*$ defines a local system of groups $\mathbb{C}^*$ on the manifold M. The cohomology of M with coefficients in this local system coincides with $H^*_{\rho_{\lambda\omega}}(M, \mathbb{C})$.

Now we can assume that ω is a Morse 1-form, i.e. in a neighbourhood of any point $\omega = \mathrm{d}f$, where f is a Morse function. In other words ω gives a multi-valued Morse function. The zeros of ω are isolated, and one can define the index of each zero. The number of zeros of ω of index p is denoted by $m_p(\omega)$.

Following Witten's scheme Pazhitnov showed in [4] that for sufficiently large real numbers λ

$$m_p(\omega) \geq b_p(\lambda, \omega).$$

11.6 Solvmanifolds and Left-Invariant Forms

A solvmanifold (nilmanifold) M is a compact homogeneous space of the form G/Γ, where G is a simply connected solvable (nilpotent) Lie group and Γ is a lattice in G [23].

Let us consider some examples of solvmanifolds (the first two of them are nilmanifolds):

1. An n-dimensional torus $T^n = \mathbb{R}^n/\mathbb{Z}^n$.
2. The Heisenberg manifold $M_3 = \mathcal{H}_3/\Gamma_3$, where $\mathcal{H}_3$ is the group of all matrices of the form

$$\begin{pmatrix} 1 & x & z \\ 0 & 1 & y \\ 0 & 0 & 1 \end{pmatrix}, \quad x, y, z \in \mathbb{R},$$

and a lattice Γ_3 is a subgroup of matrices with integer entries $x, y, z \in \mathbb{Z}$.

$$e_1 = \begin{pmatrix} 0 & 1 & 0 \\ 0 & 0 & 0 \\ 0 & 0 & 0 \end{pmatrix}, \quad e_2 = \begin{pmatrix} 0 & 0 & 0 \\ 0 & 0 & 1 \\ 0 & 0 & 0 \end{pmatrix}, \quad e_3 = \begin{pmatrix} 0 & 0 & 1 \\ 0 & 0 & 0 \\ 0 & 0 & 0 \end{pmatrix},$$

and the only one non-trivial structure relation: $[e_1, e_2] = e_3$. The left invariant 1-forms on $\mathcal{H}_3$

$$e^1 = \mathrm{d}x, \quad e^2 = \mathrm{d}y, \quad e^3 = \mathrm{d}z - x\mathrm{d}y \tag{11.17}$$

are dual to e_1, e_2, e_3 and

$$\mathrm{d}e^1 = 0, \quad \mathrm{d}e^2 = 0, \quad \mathrm{d}e^3 = \mathrm{d}(\mathrm{d}z - x\mathrm{d}y) = -\mathrm{d}x \wedge \mathrm{d}y = -e^1 \wedge e^2. \tag{11.18}$$

Now we are going to consider examples of solvmanifolds that are not nilmanifolds.

3. Let G_1 be a solvable Lie group of matrices

$$\begin{pmatrix} e^{kz} & 0 & 0 & x \\ 0 & e^{-kz} & 0 & y \\ 0 & 0 & 1 & z \\ 0 & 0 & 0 & 1 \end{pmatrix}, \tag{11.19}$$

where $e^k + e^{-k} = n \in \mathbb{N}, k \neq 0$.

G_1 can be regarded as a semi-direct product $G_1 = \mathbb{R} \ltimes \mathbb{R}^2$, where $\mathbb{R}$ acts on $\mathbb{R}^2$ (with coordinates x, y) via

$$z \to \phi(z) = \begin{pmatrix} e^{kz} & 0 \\ 0 & e^{-kz} \end{pmatrix}.$$

A lattice Γ_1 in G_1 is generated by the following matrices:

$$\begin{pmatrix} e^{k} & 0 & 0 & 0 \\ 0 & e^{-k} & 0 & 0 \\ 0 & 0 & 1 & 1 \\ 0 & 0 & 0 & 1 \end{pmatrix}, \quad \begin{pmatrix} 1 & 0 & 0 & u_1 \\ 0 & 1 & 0 & v_1 \\ 0 & 0 & 1 & 0 \\ 0 & 0 & 0 & 1 \end{pmatrix}, \quad \begin{pmatrix} 1 & 0 & 0 & u_2 \\ 0 & 1 & 0 & v_2 \\ 0 & 0 & 1 & 0 \\ 0 & 0 & 0 & 1 \end{pmatrix},$$

where $\begin{vmatrix} u_1 & v_1 \\ u_2 & v_2 \end{vmatrix} \neq 0$.

The corresponding Lie algebra $\mathfrak{g}_1$ has the following basis:

$$e_1 = \begin{pmatrix} k & 0 & 0 & 0 \\ 0 & -k & 0 & 0 \\ 0 & 0 & 0 & 1 \\ 0 & 0 & 0 & 0 \end{pmatrix}, \quad e_2 = \begin{pmatrix} 0 & 0 & 0 & 1 \\ 0 & 0 & 0 & 0 \\ 0 & 0 & 0 & 0 \\ 0 & 0 & 0 & 0 \end{pmatrix}, \quad e_3 = \begin{pmatrix} 0 & 0 & 0 & 0 \\ 0 & 0 & 0 & 1 \\ 0 & 0 & 0 & 0 \\ 0 & 0 & 0 & 0 \end{pmatrix},$$

and the following structure relations:

$$[e_1, e_2] = ke_2, \quad [e_1, e_3] = -ke_3, \quad [e_2, e_3] = 0.$$

The left-invariant 1-forms

$$e^1 = dz, \quad e^2 = e^{-kz} dx, \quad e^3 = e^{kz} dy \tag{11.20}$$

are the dual basis to e_1, e_2, e_3 and

$$de^1 = 0, \quad de^2 = -ke^{-kz} dz \wedge dx = -ke^1 \wedge e^2, \quad de^3 = ke^1 \wedge e^3. \tag{11.21}$$

As the solvable Lie group G is simply connected the fundamental group $\pi_1(G/\Gamma)$ is naturally isomorphic to the lattice Γ: $\pi_1(G/\Gamma) \cong \Gamma$.

The Lie algebra $\mathfrak{g}_1$ of G_1 considered earlier is an example of completely solvable Lie algebra. A Lie algebra $\mathfrak{g}$ is called completely solvable if $\forall X \in \mathfrak{g}$ operator $\mathrm{ad}(X)$ has only real eigenvalues.

Let G/Γ be a solvmanifold. One can identify its de Rham complex $\Lambda^*(G/\Gamma)$ with the subcomplex in $\Lambda^*(G)$

$$\Lambda^*_{\Gamma\text{inv}}(G) \subset \Lambda^*(G)$$

of left-invariant forms on G with respect to the action of the lattice Γ.

The sub-complex $\Lambda^*_{\Gamma\text{inv}}(G)$ contains in its turn the subcomplex $\Lambda^*_{\Gamma\text{inv}}(G)$ of left-invariant forms with respect to the action of G.

Taking left-invariant vector fields $X_1, \ldots, X_{p+1}$ and a left-invariant p-form $\xi \in \Lambda^*_{G\text{inv}}(G)$ in the formula (11.12) we have:

$$\mathrm{d}\xi(X_1, \ldots, X_{p+1}) = \sum_{1 \le i < j \le p+1} (-1)^{i+j} \xi([X_i, X_j], X_1, \ldots, \hat{X}_i, \ldots, \hat{X}_j, \ldots, X_{p+1}). \tag{11.22}$$

The Lie algebra of left-invariant vector fields on G is naturally isomorphic to the tangent Lie algebra $\mathfrak{g}$. Hence one can identify the space $\Lambda^p_{G\text{inv}}(G)$ with the space $\Lambda^p(\mathfrak{g}^*)$ of skew-symmetric polylinear functions on $\mathfrak{g}$.

The differential d defined by (11.22) provides us with the cochain complex of the Lie algebra $\mathfrak{g}$:

$$\mathbb{R} \xrightarrow{\mathrm{d}_0 = 0} \mathfrak{g}^* \xrightarrow{\mathrm{d}} \Lambda^2(\mathfrak{g}^*) \xrightarrow{\mathrm{d}} \Lambda^3(\mathfrak{g}^*) \xrightarrow{\mathrm{d}} \cdots \tag{11.23}$$

The dual of the Lie bracket $[,] : \Lambda^2(\mathfrak{g}) \to \mathfrak{g}$ gives a linear mapping

$$\delta : \mathfrak{g}^* \to \Lambda^2(\mathfrak{g}^*).$$

Consider a basis $e_1, \ldots, e_n$ of $\mathfrak{g}$ and its dual basis $e^1, \ldots, e^n$. Then we have the following relation:

$$\mathrm{d}e^k = -\delta e^k = -\sum_{i<j} C^k_{ij} \mathrm{d}e^i \wedge \mathrm{d}e^j, \tag{11.24}$$

where $[e_i, e_j] = \sum C^k_{ij} e_k$. The differential d is completely determined by (11.24) and the following property:

$$\mathrm{d}(\xi_1 \wedge \xi_2) = \mathrm{d}\xi_1 \wedge \xi_2 + (-1)^{\deg \xi_1} \xi_1 \wedge \mathrm{d}\xi_2, \ \forall \xi_1, \xi_2 \in \Lambda^*(\mathfrak{g}^*).$$

Cohomology of the complex $(\Lambda^*(\mathfrak{g}^*), \delta)$ is called the *cohomology (with trivial coefficients) of the Lie algebra* $\mathfrak{g}$ and is denoted by $H^*(\mathfrak{g})$.

Let us consider the inclusion

$$\psi : \Lambda^*(\mathfrak{g}) \to \Lambda^*(G/\Gamma).$$

Let G/Γ be a compact solvmanifold, where G is a completely solvable Lie group, then $\psi : \Lambda^*(\mathfrak{g}) \to \Lambda^*(G/\Gamma)$ induces the isomorphism $\psi^* : H^*(\mathfrak{g}) \to H^*(G/\Gamma, \mathbb{R})$ in cohomology (Hattori's theorem [12], Nomizu's theorem for nilmanifolds [8]).

Let us return to our examples:

1. The cohomology classes $H^*(\mathbb{T}^n, \mathbb{R})$ are represented by invariant forms
$$\mathrm{d}x^{i_1} \wedge \cdots \wedge \mathrm{d}x^{i_q}, \quad 1 \le i_1 < \cdots < i_q \le n,\ q = 1, \ldots, n.$$
2. $H^*(\mathcal{H}_3/\Gamma_3, \mathbb{R})$ is spanned by the cohomology classes of the following left-invariant forms:
$$\mathrm{d}x,\ \mathrm{d}y,\ \mathrm{d}y \wedge \mathrm{d}z,\ \mathrm{d}x \wedge (\mathrm{d}z - x\mathrm{d}y),\ \mathrm{d}x \wedge \mathrm{d}y \wedge \mathrm{d}z.$$
3. $H^*(G_1/\Gamma_1, \mathbb{R})$ is spanned by the cohomology classes of:
$$e^1 = \mathrm{d}z,\ e^2 \wedge e^3 = \mathrm{d}x \wedge \mathrm{d}y,\ e^1 \wedge e^2 \wedge e^3 = \mathrm{d}x \wedge \mathrm{d}y \wedge \mathrm{d}z.$$

11.7 Deformed Differential and Lie Algebra Cohomology

From the definition of Lie algebra cohomology it follows that $H^1(\mathfrak{g})$ is the dual space to $\mathfrak{g}/[\mathfrak{g}, \mathfrak{g}]$:

1. $b^1(\mathfrak{g}) = \dim H^1(\mathfrak{g}) \ge 2$ for a nilpotent Lie algebra $\mathfrak{g}$ (Dixmier's theorem [15]);
2. $b^1(\mathfrak{g}) \ge 1$ for a solvable Lie algebra $\mathfrak{g}$;
3. $b^1(\mathfrak{g}) = 0$ for a semi-simple Lie algebra $\mathfrak{g}$.

Consider a Lie algebra $\mathfrak{g}$ with a non-trivial $H^1(\mathfrak{g})$. Let $\omega \in \mathfrak{g}^*, \mathrm{d}\omega = 0$. One can define:

1. A new deformed differential d_ω in $\Lambda^*(\mathfrak{g}^*)$ by the formula
$$\mathrm{d}_\omega(a) = \mathrm{d}a + \omega \wedge a.$$
2. A one-dimensional representation
$$\rho_\omega : \mathfrak{g} \to \mathbb{K},\ \rho_\omega(\xi) = \omega(\xi), \xi \in \mathfrak{g}.$$

Now we recall the definition of Lie algebra cohomology associated with a representation. Let $\mathfrak{g}$ be a Lie algebra and $\rho : \mathfrak{g} \to \mathfrak{gl}(V)$ its linear representation. We denote by $C^q(\mathfrak{g}, V)$ the space of q-linear alternating mappings of $\mathfrak{g}$ into V. Then one can consider an algebraic complex:
$$V = C^0(\mathfrak{g}, V) \xrightarrow{\mathrm{d}} C^1(\mathfrak{g}, V) \xrightarrow{\mathrm{d}} C^2(\mathfrak{g}, V) \xrightarrow{\mathrm{d}} C^3(\mathfrak{g}, V) \xrightarrow{\mathrm{d}} \cdots$$
where the differential d is defined by:
$$(\mathrm{d}f)(X_1, \ldots, X_{q+1}) = \sum_{i=1}^{q+1} (-1)^{i+1} \rho(X_i)(f(X_1, \ldots, \hat{X}_i, \ldots, X_{q+1})) \\ + \sum_{1 \le i < j \le q+1} (-1)^{i+j-1} f([X_i, X_j], X_1, \ldots, \hat{X}_i, \ldots, \hat{X}_j, \ldots, X_{q+1}). \quad (11.25)$$

The cohomology of the complex $(C^*(\mathfrak{g}, V), \mathrm{d})$ is called *the cohomology of the Lie algebra* $\mathfrak{g}$ associated to the representation $\rho : \mathfrak{g} \to \mathfrak{gl}(V)$.

Let $\mathfrak{g}$ be a Lie algebra and $\omega \in \mathfrak{g}^*$ is a closed 1-form. Then the complex $(\Lambda^*(\mathfrak{g}^*), \mathrm{d}_\omega)$ coincides with the cochain complex of the Lie algebra $\mathfrak{g}$ associated with the one-dimensional representation $\rho_\omega : \mathfrak{g} \to \mathbb{K}$, where $\rho_\omega(\xi) = \omega(\xi)$, $\xi \in \mathfrak{g}$.

The proof follows from the formula:

$$(\omega \wedge a)(X_1, \dots, X_{q+1}) = \sum_{i=1}^{q+1} (-1)^{i+1} \omega(X_i)(a(X_1, \dots, \hat{X}_i, \dots, X_{q+1})).$$

The cohomology $H^*_\omega(\mathfrak{g})$ coincides with the Lie algebra cohomology with trivial coefficients if $\omega = 0$. If $\omega \neq 0$ the deformed differential d_ω is not compatible with the exterior product $\wedge$ in $\Lambda^*(\mathfrak{g})$

$$\mathrm{d}_\omega(a \wedge b) = \mathrm{d}(a \wedge b) + \omega \wedge a \wedge b \neq \mathrm{d}_\omega(a) \wedge b + (-1)^{\mathrm{deg} a} a \wedge \mathrm{d}_\omega(b)$$

and the cohomology $H^*_\omega(\mathfrak{g})$ has no natural multiplicative structure.

Let G/Γ be a compact solvmanifold, where G is a completely solvable Lie group and $\tilde{\omega}$ is a closed 1-form on G/Γ. From the previous sections it follows that the cohomology $H^*_{\tilde{\omega}}(G/\Gamma, \mathbb{C})$ is isomorphic to the Lie algebra cohomology $H^*_\omega(\mathfrak{g})$ where $\omega \in \mathfrak{g}^*$ is the left-invariant 1-form that represents the class $[\tilde{\omega}] \in H^1(G/\Gamma, \mathbb{R})$.

One can define by means of ω a one-dimensional representation $\rho_\omega : G \to \mathbb{C}^*$:

$$\rho_\omega(g) = \exp \int_{\gamma(e,g)} \omega,$$

where $\gamma(e, g)$ is a path connecting the identity e with $g \in G$ (let us recall that G is a simply connected). As ω is the left invariant 1-form then

$$\int_{\gamma(e,g_1 g_2)} \omega = \int_{\gamma(e,g_1)} \omega + \int_{\gamma(g_1, g_1 g_2)} \omega = \int_{\gamma(e,g_1)} \omega + \int_{g_1^{-1}\gamma(e,g_2)} \omega$$

holds on and $\rho_\omega(g_1 g_2) = \rho_\omega(g_1)\rho_\omega(g_2)$.

The representation ρ_ω induces the representation of corresponding Lie algebra $\mathfrak{g}$ (we denote it by the same symbol): $\rho_\omega(X) = \omega(X)$.

Let $\mathfrak{g}$ be an n-dimensional real completely solvable Lie algebra (or complex solvable) and $b^1(\mathfrak{g}) = \dim H^1(\mathfrak{g}) = k \geq 1$. Then exists a basis $e^1, \dots, e^n$ in $\mathfrak{g}^*$ such that

$$\begin{gathered} \mathrm{d}e^1 = \dots = \mathrm{d}e^k = 0, \\ \mathrm{d}e^{k+s} = \alpha_{k+s} \wedge e^{k+s} + P_{k+s}(e^1, \dots, e^{k+s-1}), \quad s = 1, \dots, n-k, \end{gathered} \tag{11.26}$$

where

$$\begin{aligned} \alpha_{k+s} &= \alpha_{s;1}e^1 + \alpha_{s;2}e^2 + \cdots + \alpha_{s;k}e^k, \\ P_{k+s}(e^1, \ldots, e^{k+s-1}) &= \sum_{1 \le i < j \le k+s-1} P_{s;i,j} e^i \wedge e^j. \end{aligned} \tag{11.27}$$

It is convenient to define $\alpha_i = 0, i = 1, \ldots, k$. The set $\{\alpha_1, \ldots, \alpha_n\}$ of closed 1-forms is in fact the set of the weights of completely reducible representation associated to the adjoint representation $X \to \mathrm{ad}(X)$.

For the proof we apply Lie's theorem to the adjoint representation ad restricted to the commutant $[\mathfrak{g}, \mathfrak{g}]$:

$$X \in \mathfrak{g} \to \mathrm{ad}(X) : [\mathfrak{g}, \mathfrak{g}] \to [\mathfrak{g}, \mathfrak{g}].$$

Namely we can choose a basis $e_{k+1}, \ldots, e_n$ in $[\mathfrak{g}, \mathfrak{g}]$ such that the subspaces $V_i, i = k+1, \ldots, n$ spanned by $e_i, \ldots, e_n$ are invariant with respect to the representation ad. Then we add $e_1, \ldots, e_k$ in order to get a basis of the whole $\mathfrak{g}$. For the forms of the dual basis $e^1, \ldots, e^n$ in $\mathfrak{g}^*$ we have formulas (11.26).

Let us consider a new canonical basis of $\mathfrak{g}^*$:

$$\begin{aligned} \tilde{e}^1 &= e^1, \ldots, \tilde{e}^k = e^k, \\ \tilde{e}^{k+s} &= t^{2(s-1)} e^{k+s}, \quad s = 1, \ldots, n-k, \end{aligned} \tag{11.28}$$

where $t > 0$ is a real parameter.

Then for the differential d_ω in the complex $\Lambda^*(\tilde{e}^1, \ldots, \tilde{e}^n)$ we have:

$$\mathrm{d}_\omega = \mathrm{d}_0 + \omega \wedge + t d_1 + t^2 \mathrm{d}_2 + \cdots, \quad \mathrm{d}_0 \tilde{e}^i = \alpha_i \wedge \tilde{e}^i.$$

In particular

$$(\mathrm{d}_0 + \omega \wedge)(\tilde{e}^{i_1} \wedge \cdots \wedge \tilde{e}^{i_q}) = (\alpha_{i_1} + \cdots + \alpha_{i_q} + \omega) \wedge \tilde{e}^{i_1} \wedge \cdots \wedge \tilde{e}^{i_q}.$$

Now one can define the scalar product in $\Lambda^q(\tilde{e}^1, \ldots, \tilde{e}^n)$ declaring the set $\{e^{i_1} \wedge \cdots \wedge e^{i_q}\}$ of basic q-forms as an orthonormal basis of $\Lambda^q(\tilde{e}^1, \ldots, \tilde{e}^n)$. Then

$$\begin{aligned} \mathrm{d}_\omega^* \mathrm{d}_\omega + \mathrm{d}_\omega \mathrm{d}_\omega^* &= R_0 + t R_1 + t^2 R_2 + \cdots, \\ R_0(\tilde{e}^{i_1} \wedge \cdots \wedge \tilde{e}^{i_q}) &= \|\alpha_{i_1} + \cdots + \alpha_{i_q} + \omega\|^2 \tilde{e}^{i_1} \wedge \cdots \wedge \tilde{e}^{i_q}. \end{aligned} \tag{11.29}$$

As $t \to 0$ the minimal eigenvalue of $\mathrm{d}_\omega^* \mathrm{d}_\omega + \mathrm{d}_\omega \mathrm{d}_\omega^*$ converges to the minimal eigenvalue of R_0. Thus if

$$\alpha_{i_1} + \cdots + \alpha_{i_q} + \omega \neq 0, \quad 1 \le i_1 < i_2 < \cdots < i_q \le n$$

then $H^q_\omega(\mathfrak{g}) = 0$ (Fig. 11.2).

Recall that $\alpha_1 = \cdots = \alpha_k = 0$ and let us introduce the finite subset $\Omega_\mathfrak{g} \subset H^1(\mathfrak{g})$ such that:

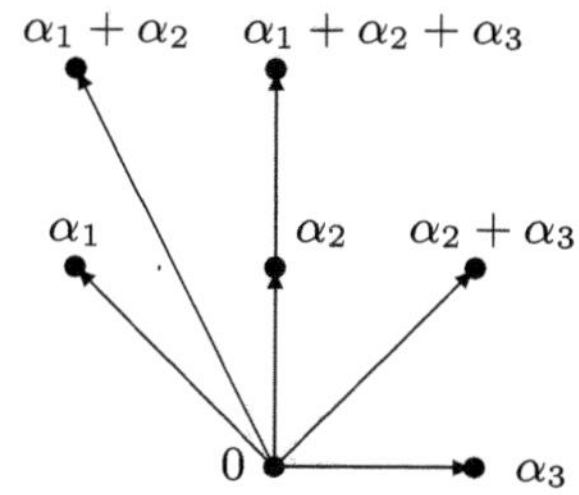

Fig. 11.2. The finite subset $\Omega_{\mathfrak{g}} \subset H^1(\mathfrak{g})$

$$\Omega_{\mathfrak{g}} = \{\alpha_{i_1}+\dots+\alpha_{i_s} | 1 \le i_1<\dots<i_s \le n, s=1,\dots,n\}. \tag{11.30}$$

It follows that if

$$-\omega \notin \Omega_{\mathfrak{g}}$$

then the total cohomology $H^*_\omega(\mathfrak{g})$ is trivial: $H^*_\omega(\mathfrak{g}) \equiv 0$.

One can easily remark that the subset $\Omega_{\mathfrak{g}}$ is well defined and does not depend on the ordering of weights α_i.

Let G/Γ be a compact solvmanifold, where G is a completely solvable Lie group. Then the left-invariant closed 1-forms from $\Omega_{\mathfrak{g}}$ define a finite subset in $H^1(G/\Gamma, \mathbb{R})$. We denote this subset by $\Omega_{G/\Gamma}$. Let ω be a closed 1-form on G/Γ. If the cohomology class

$$-[\omega] \notin \Omega_{G/\Gamma}$$

then the total cohomology $H^*_\omega(G/\Gamma, \mathbb{R})$ is trivial: $H^*_\omega(G/\Gamma, \mathbb{R}) \equiv 0$. The subset $\Omega_{G/\Gamma}$ is well defined in terms of the corresponding Lie algebra $\mathfrak{g}$. The corresponding Lie algebra $\mathfrak{g}$ must to be unimodular, i.e. the left-invariant n-form $e^1 \wedge \dots \wedge e^n$ determines non-exact volume form on G/Γ and hence

$$\alpha_1+\alpha_2+\dots+\alpha_n = 0.$$

If G/Γ is a compact nilmanifold then all the weights $\alpha_i, i=1,\dots,n$ are trivial and therefore $\Omega_{G/\Gamma} = \{0\}$. Hence the cohomology $H^*_\omega(G/\Gamma, \mathbb{R})$ of a nilmanifold G/Γ is trivial if and only if the form ω is non-exact.

Let us consider a 3-dimensional solvmanifold G_1/Γ_1 defined in Sect. 11.6. We recall that the corresponding Lie algebra $\mathfrak{g}_1$ is defined by its basis e_1, e_2, e_3 and the following non-trivial brackets:

$$[e_1, e_2] = ke_2, \quad [e_1, e_3] = -ke_3.$$

For the dual basis of left-invariant 1-forms $e^1 = \mathrm{d}z, e^2 = e^{-kz}\mathrm{d}x, e^3 = e^{kz}\mathrm{d}y$ we had

$$\mathrm{d}e^1 = 0,\ \mathrm{d}e^2 = -ke^1 \wedge e^2,\ \mathrm{d}e^3 = ke^1 \wedge e^3.$$

Hence $\alpha_1 = 0, \alpha_2 = -ke^1, \alpha_3 = ke^1$ and $\alpha_2+\alpha_3 = 0$ (Fig. 11.3).

So it is easy to see that

$$\Omega_{G_1/\Gamma_1} = \{\pm k[e^1]\}$$

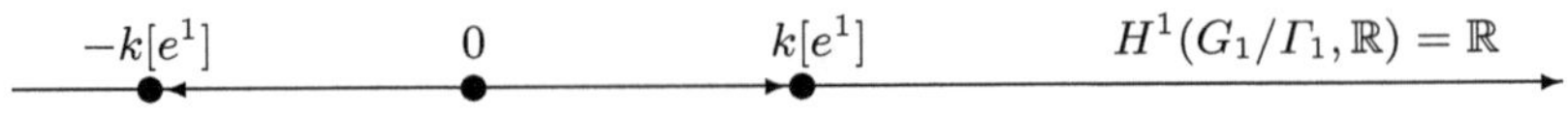

Fig. 11.3. The finite subset Ω_{G_1/Γ_1}

and therefore the cohomology $H^*_\omega(G_1/\Gamma_1, \mathbb{R})$ is trivial if $[\omega] \neq 0, \pm k[e^1]$.

(a) $H^*_{k[e^1]}(G_1/\Gamma_1, \mathbb{R})$ is spanned by two classes:

$$e^2 = e^{-kz}dx, \quad e^1 \wedge e^2 = dz \wedge e^{-kz}dx.$$

(b) $H^*_{-k[e^1]}(G_1/\Gamma_1, \mathbb{R})$ is spanned by two classes:

$$e^3 = e^{kz}dy, \quad e^1 \wedge e^3 = dz \wedge e^{kz}dy.$$

Hence we have the following Betti numbers $b^p_\omega = \dim H^p_\omega(G_1/\Gamma_1, \mathbb{R})$ of the solvmanifold G_1/Γ_1:

$$\begin{aligned} 1.\quad & b^0_{\pm ke^1} = 0, b^1_{\pm ke^1} = b^2_{\pm ke^1} = 1, b^3_{\pm ke^1} = 0. \\ 2.\quad & b^0_0 = b^1_0 = b^2_0 = b^3_0 = 1. \end{aligned} \tag{11.31}$$

It was proved by Mostow in [20] that any compact solvmanifold G/Γ is a bundle with toroid as base space and nilmanifold as fibre, in particular a solvmanifold G/Γ is fibred over the circle $\pi : G/\Gamma \to S^1$. Hence the 1-form $\pi^*(d\varphi)$ on G/Γ has no critical points: $m_p(\pi^*(d\varphi)) = 0,\ \forall p$. It follows from Pajitnov's theorem [4] that for λ sufficiently large we have $H^p_{\lambda\pi^*(d\varphi)}(G/\Gamma, \mathbb{R}) = 0,\ \forall p$.

Now we are going to introduce an example of solvmanifold G/Γ with non-completely solvable Lie group G (see [21]). Let G_2 be a solvable Lie group of matrices

$$\begin{pmatrix} \cos 2\pi z & \sin 2\pi z & 0 & x \\ -\sin 2\pi z & \cos 2\pi z & 0 & y \\ 0 & 0 & 1 & z \\ 0 & 0 & 0 & 1 \end{pmatrix}. \tag{11.32}$$

A lattice Γ_2 in G_2 is generated by the following matrices:

$$\begin{pmatrix} \cos \frac{2\pi n}{p} & \sin \frac{2\pi n}{p} & 0 & 0 \\ -\sin \frac{2\pi n}{p} & \cos \frac{2\pi n}{p} & 0 & 0 \\ 0 & 0 & 1 & \frac{n}{p} \\ 0 & 0 & 0 & 1 \end{pmatrix}, \quad \begin{pmatrix} 1 & 0 & 0 & u_1 \\ 0 & 1 & 0 & v_1 \\ 0 & 0 & 1 & 0 \\ 0 & 0 & 0 & 1 \end{pmatrix}, \quad \begin{pmatrix} 1 & 0 & 0 & u_2 \\ 0 & 1 & 0 & v_2 \\ 0 & 0 & 1 & 0 \\ 0 & 0 & 0 & 1 \end{pmatrix},$$

where n is an integer, $p = 2, 3, 4, 6$ and $\begin{vmatrix} u_1 & v_1 \\ u_2 & v_2 \end{vmatrix} \neq 0$, or another type: $\tilde{\Gamma}_2$ is generated by the following matrices:

$$\begin{pmatrix} 1 & 0 & 0 & 1 \\ 0 & 1 & 0 & 0 \\ 0 & 0 & 1 & 0 \\ 0 & 0 & 0 & 1 \end{pmatrix}, \quad \begin{pmatrix} 1 & 0 & 0 & 0 \\ 0 & 1 & 0 & 1 \\ 0 & 0 & 1 & 0 \\ 0 & 0 & 0 & 1 \end{pmatrix}, \quad \begin{pmatrix} 1 & 0 & 0 & u \\ 0 & 1 & 0 & v \\ 0 & 0 & 1 & n \\ 0 & 0 & 0 & 1 \end{pmatrix},$$

where n is an integer. The corresponding Lie algebra $\mathfrak{g}_2$ has the following basis:

$$e_1 = \begin{pmatrix} 0 & 2\pi & 0 & 0 \\ -2\pi & 0 & 0 & 0 \\ 0 & 0 & 0 & 1 \\ 0 & 0 & 0 & 0 \end{pmatrix}, \quad e_2 = \begin{pmatrix} 0&0&0&1 \\ 0&0&0&0 \\ 0&0&0&0 \\ 0&0&0&0 \end{pmatrix}, \quad e_3 = \begin{pmatrix} 0&0&0&0 \\ 0&0&0&1 \\ 0&0&0&0 \\ 0&0&0&0 \end{pmatrix},$$

and the following structure relations:

$$[e_1, e_2] = -2\pi e_3, \quad [e_1, e_3] = 2\pi e_2, \quad [e_2, e_3] = 0.$$

As the eigenvalues of $\mathrm{ad}(e_1)$ are equal to $0, \pm 2\pi i$ the Lie group G_2 is not completely solvable.

The left-invariant 1-forms

$$e^1 = \mathrm{d}z, \quad e^2 = \cos 2\pi z \mathrm{d}x - \sin 2\pi z \mathrm{d}y, \quad e^3 = \sin 2\pi z \mathrm{d}x + \cos 2\pi z \mathrm{d}y \qquad (11.33)$$

are the dual basis to e_1, e_2, e_3 and

$$\mathrm{d}e^1 = 0, \quad \mathrm{d}e^2 = -2\pi e^1 \wedge e^3, \quad \mathrm{d}e^3 = 2\pi e^1 \wedge e^2. \qquad (11.34)$$

The cohomology $H^*(\mathfrak{g}_2)$ is spanned by the cohomology classes of:

$$e^1, \; e^2 \wedge e^3, \; e^1 \wedge e^2 \wedge e^3.$$

But

$$\dim H^1(\mathfrak{g}_2) = 1 \neq \dim H^1(G_2/\Gamma_2, \mathbb{R}) = 3.$$

This example shows that, generally speaking, Hattori's theorem does not hold for non-completely solvable Lie groups G, but the inclusion of left-invariant differential forms $\psi : \Lambda^*(\mathfrak{g}^*) \to \Lambda^*(G/\Gamma)$ always induces the injection ψ^* in cohomology.

References

1. S.P. Novikov, Soviet Math. Dokl. **24**, 222–226, (1981)
2. S. P. Novikov, Russ. Math. Surveys **37** (5), 1–56, (1982)
3. S. P. Novikov, Soviet Math. Dokl. **33** (5), 551–555, (1986)
4. A.V. Pazhitnov, Soviet Math. Dokl. **35**, 1–2, (1987)
5. E. Witten, *Supersymmetry and Morse theory*, J. Differential Geom. **17**, 661–692, (1982)
6. S. P. Novikov, *On the exotic De-Rham cohomology. Perturbation theory as a spectral sequence*, arXiv:math-ph/0201019
7. L. Alaniya, Russ. Math. Surveys **54** (5) 1019–1020 (1999)
8. K. Nomizu, Ann. Math. **59**, 531–538, (1954)
9. D.V. Millionshchikov, Russ. Math. Surveys **57** (4) 813–814, (2002)
10. D.V. Millionshchikov, Math. Notes (in Russian) 77(1–2), 61–71 (2005)

11. J. Dixmier, Acta Sci. Math. (Szeged), **16** (4), 226–250, (1955)
12. A. Hattori, J. Fac. Sci. Univ. Tokyo, Sect. 1, **8** (4), 289–331 (1960)
13. P.A.M. Dirac, Phys. Rev. **74**, 817, (1948)
14. Y. Aharonov, D. Bohm, Phys. Rev. **115**(3) 145, (1959)
15. Y. Aharonov, D. Bohm, Phys. Rev. **123** (4) 1511, (1961)
16. P.A. Horvathy, *Classical Action, the Wu–Yang Phase Factor and Prequantization*, in Proc. conf., Aix-en-Provance, Salamanca. Lecture notes in mathematics, vol. 836 (Springer, Berlin Heidelberg New York, 1980), pp. 67–90. 1980
17. T.T. Wu, C.N. Yang, Phys. Rev. **D12**, 3845, (1975)
18. I.A. Dynnikov, *Semiclassical motion of the electron. A proof of the Novikov conjecture in general position and counterexamples*, Solitons, Geometry and Topology: on the Crossroad (V.M. Buchstaber, ed.), Am. Math. Soc. Transl. **179**(2), 45–73 (1997)
19. I.A. Dynnikov, Russ. Math. Surveys **54** (1) 21–60, (1999)
20. G.D. Mostow, Ann. Math. **73**, 20–48, (1961)
21. L. Ausllander, L. Green, F. Hahn, Ann. Math. Stud. **53**, (Princeton University Press Princeton, NJ, 1963) p. 107
22. J.W. Milnor, *Morse theory*, (Princeton University Press, Princeton, NJ 1963)
23. S. Raghunathan, *Discrete subgroups of Lie groups*, (Springer, Berlin Heidelberg New York) 1972

12

The Spectral Geometry of Riemann Surfaces

R. Brooks

Summary. This chapter is spread out over a number of papers and also builds on my earlier work on the relationship between the spectral geometry of manifolds and the spectral geometry of graphs. It seemed to be a reasonable idea to put together these ideas in one overall framework, which will be accessible to someone at the graduate level.

The material naturally breaks up into a number of areas, each one having connections to basic graduate material, but putting these different pieces together demands a fair amount of breadth. We hope to supply this breadth in the chapters.

12.1 Introduction

These are notes to accompany my lectures at the Institut Henri Poincaré. The idea of these lectures is to present the approach of myself and Eran Makover toward understanding the spectral geometry of a typical Riemann surface.

This work is spread out over a number of papers and also builds on my earlier work on the relationship between the spectral geometry of manifolds and the spectral geometry of graphs. It seemed to be a reasonable idea to put together these ideas in one overall framework, which will be accessible to someone at the graduate level.

Unfortunately, this is not the approach that we will take in these notes. The material naturally breaks up into a number of areas, each one having connections to basic graduate material, but putting these different pieces together demands a fair amount of breadth. We hope to supply this breadth in the lectures.

Our hope in these notes is somewhat more modest. Each section of the notes will be devoted to a section of the material. Our plan is to make each section pretty much independent, so that someone can pick up a particular topic. The task of knitting the different pieces together to get a coherent overall picture will have to wait, perhaps for a long time. In the meantime, it is hoped that the various sections will appear in a manner that will allow the students to keep up with the lectures.

Our expectation is that the background of the students in complex analysis and Riemann surfaces is perhaps the weakest point. For that purpose, the primary purpose of the notes will be to supply the reader with the necessary background in this area. Another area that we expect the students to be relatively weak is in probability theory. As this was the part of the material that came least naturally to us, we will also try to be fairly explicit here. Our plan is to post these notes on our web site as well as on the web site of the IHP, in a timely manner as the notes are written. We encourage the reader to check for updates and additions.

It is a pleasure to thank Thierry Coulhon and the organizers of the special trimester "Noyaux de Chaleur" for the invitation to participate in the special semester and to deliver these lectures.

12.2 An Opening Question

It has long been an interest of mine to pass between graphs and manifolds. The standard picture is in some sense quite clear and elegant. Nonetheless, it seems to me that there is much that can be added to the standard picture.

To give you an idea of the problem, let us consider a family of graphs $X^{p,q}$ considered by Lubotsky, Phillips, and Sarnak. They are indexed by two prime number p and q, and are $p+1$-regular graphs that are Cayley graphs (with respect to some nice choice of generators) of either $PSL(2,\mathbb{Z}/q)$ or $PGL(2,\mathbb{Z}/q)$, depending on certain properties of p and q. They have the property that their first eigenvalue is as large as possible, and for that reason were called *Ramanujan graphs* by L-P-S.

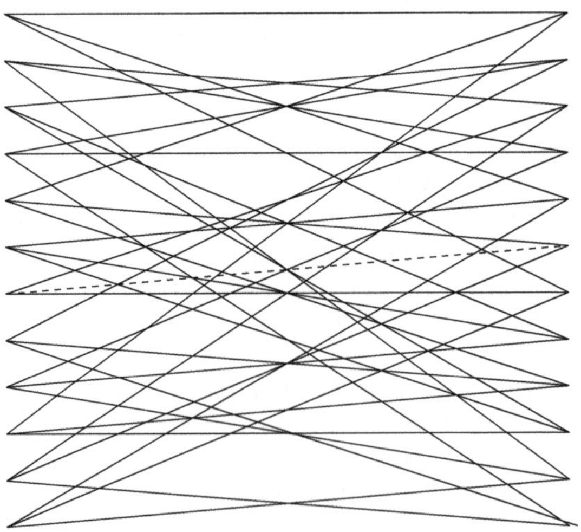

Fig. 12.1. The graph X^{23} according to Sarnak

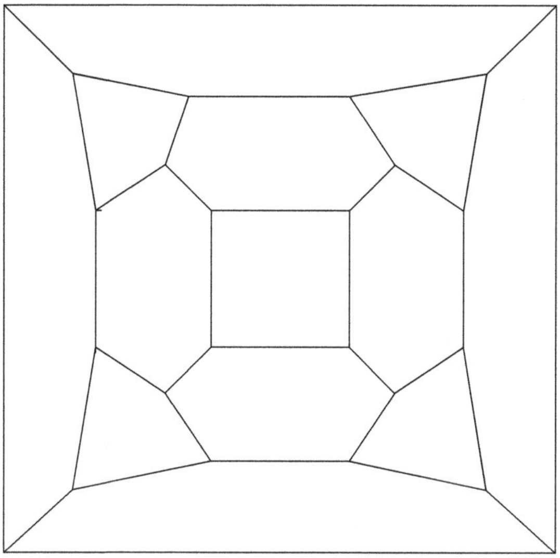

Fig. 12.2. The graph $X^{2,3}$ according to Brooks and Zuk

The simplest of them is the graph X^{23} (Fig. 12.1). Here is a picture of $X^{2,3}$ taken from the book of Sarnak:

I think that you will agree that it is hard to say anything intelligent about this graph from the picture. Indeed, Sarnak seems to have overlooked the fact that the graph as he drew it was not even three-regular. We have supplied the edge that Sarnak missed by writing it as a dotted line. While playing around with some of the ideas which will follow, Andrzej Zuk and I noticed that one can rearrange the graph in a way that is easier to understand. It is shown below (Fig. 12.2).

In fact, all of the Ramanujan graphs have a nice structure that makes them come out nice – not, perhaps, so nice as in this example, but still one that suggests a reasonable geometric picture. The question is: what, if anything, is this picture trying to tell us? Can we make good use of this extra structure? Does it suggest a more general picture where there is more to the geometry of graphs than meets the eye? While the investigations below came from other sources, I think that a good way to understand where we are going is by considering these questions in light of the example of the two ways of writing the same graph.

12.3 The Noncompact Case

In this section, we want to extend our theorem on the behavior of λ_1 under coverings to the case where the base manifold S is a Riemann surface of finite area:

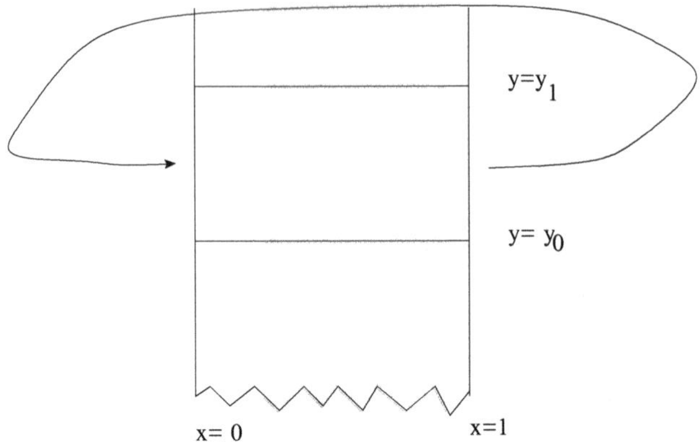

Fig. 12.3. What a cusp looks like

Theorem 1. *Let S be a Riemann surface of finite area, and let $\{S_i\}$ be a family of coverings of S given by graphs $\{\Gamma_i\}$.*
Then $\lambda_i(S_i) \to 0$ as $i \to \infty$ if and only if $h(\Gamma_i) \to 0$ as $i \to \infty$

The basic idea of the proof is a following: if C is a cusp of S, then C looks as follows:

Let S_{y_j} be $S \cap \{y : y \leq y_j\}$, where we have chosen some y_j for each cusp (Fig. 12.3). S_{y_j} is of course compact with boundary. Given our coverings S_i, we may lift S_{y_j} to S_i to get a family of coverings S_{i,y_j} of S_{y_j}.

We may now apply our proof in the compact case to this setting, where we replace λ_1 with λ_1^N, the first eigenvalue with Neumann boundary conditions, and replace the Cheeger constant $h(S)$ with the Cheeger constant $h^N(S_{i,y_j})$ with Neumann boundary conditions. We now have enough compactness to see that the original argument goes through. The only point at which we have to be careful is passing from $\lambda_1(S_i)$ to $\lambda_1^N(S_{i,y_j})$.

In general, if S^* has boundary, we have

$$\lambda_1(S^*) = inf_{\int_{S^*} f d\,\text{area}=0} \frac{\int_{S^*} \| \operatorname{grad}(f) \|^2}{\int_{S^*} f^2}.$$

Now let f be an eigenvalue of Δ on S_i with eigenvalue

$$\lambda_1 = 1/4 - s^2.$$

(If λ_1 is bigger than 1/4, there is nothing to prove.) We need to compare the Rayleigh quotient of f on S_i with the Rayleigh quotient of f on $S_{(i,y_j)}$. Clearly, the numerator is less, since we are integrating the same over a small area, but the denominator is also less, for the same reason. We want to show that the denominator is not *too much* less.

We must also worry about the fact that $\int S_{i,y_j} f$ need no longer be equal to zero. We can modify this by subtracting a constant from f, which does not change grad(f), but we must worry that this constant is not too large.

Both worries will be taken care of by showing that "not much is happening far out in the cusps," by comparing what happens there to what happens closer in. What we will show is:

Theorem 2. *Let C_{y_j} be the part of the cusp where $y > y_j$.*

Let f be an L^2 function on C with eigenvalue $\lambda = 1/4 - s^2$, and $y_1 > y_0$. Then

$$\int_{C_{y_1}} f^2 \leq \left(\frac{y_0}{y_1}\right)^{2s} \int_{C_{y_0}} f^2.$$

The idea of the proof is the following: we may fix coordinates (x, y) in C, where $0 \leq x \leq 1$ and y sufficiently large. The Laplacian of f is then given by

$$\Delta(f) = -y^2 \left(\frac{\partial^2}{\partial x^2} f + \frac{\partial^2}{\partial y^2} f\right).$$

Since f is periodic in x, we may write out its Fourier series in x as a function of y:

$$f(x, y) = \sum_n a_n(y) \cos(2\pi n x) + b_n(y) \sin(nx).$$

The equation

$$\Delta(f) = \lambda f$$

then translates to the differential equations

$$a_n''(y) = \left(4\pi^2 n^2 - \frac{\lambda}{y^2}\right) a_n,$$

where differentiation is understood with respect to y, similarly for b_n.

Putting aside the nuisance that these functions are of length 1/2 and not 1, we have that

$$\int_{C_{y_i}} f^2 = \int_{y=y_i}^{\infty} \frac{\sum_n a_n^2(y) + b_n^2(y)}{y^2} dy.$$

Let us first examine the case $n = 0$. We are then looking at the equation

$$a_0'' = \frac{-\lambda}{y^2} a_0.$$

This has solutions

$$a_0 = c_1 y^{(1/2)-s} + c_2 y^{(1/2)+s}.$$

In order for this term to be L^2, we must have $c_2 = 0$. Thus, we must have

$$a_0 = c_1 y^{(1/2)-s},$$

and it is easily seen that

$$\int_{y_1}^{\infty} a_0^2 \frac{1}{y^2}\, dy = (\mathrm{const})(y_1)^{-2s}.$$

Hence

$$\int_{y_1}^{\infty} a_0^2 \frac{1}{y^2}\, dy = \left(\frac{y_0}{y_1}\right) \int_{y_0}^{\infty} a_0^2 \frac{1}{y^2}\, dy.$$

We would like to supply a similar analysis to the other terms. The treatment of the a_n's and b_n's is exactly the same, so we will focus on the a_n's. The idea is to apply the standard techniques of Sturm–Liouville comparison to study the behavior of the solutions of the equations. We will do this in several ways.

First of all, as y gets large, the term $-\lambda/y$ becomes negligible in comparison to the terms $4\pi^2 n^2$. Thus, the differential equation for a_n has two solutions, one decaying like $e^{-2\pi n y}$ and the other blowing up like $e^{2\pi n y}$, for y large. In order to preserve L^2-ness, the second term cannot appear. Thus, there is a unique solution (up to constants) F_n of this equation, which decays like $e^{-2\pi n y}$ for y large. We may normalize F_n by insisting, say, that $F_n(y_0) = 1$.

We then compare F_n to $F_0 = y^{(1/2)} - s$. we find first of all that F_n is positive for all values, and second that F_n/F_0 is a decreasing function of y.

It follows that

$$\frac{\int_{y=y_1}^{\infty} F_n^2 y^{-2}\, dy}{\int_{y=y_1}^{\infty} F_0^2 y^{-2}\, dy} \le \frac{\int_{y=y_0}^{\infty} F_n^2 y^{-2}\, dy}{\int_{y=y_0}^{\infty} F_0^2 y^{-2}\, dy}.$$

Rewriting this as

$$\frac{\int_{y=y_1}^{\infty} F_n^2 y^{-2}\, dy}{\int_{y=y_0}^{\infty} F_n^2 y^{-2}\, dy} \le \frac{\int_{y=y_1}^{\infty} F_0^2 y^{-2}\, dy}{\int_{y=y_0}^{\infty} F_0^2 y^{-2}\, dy}.$$

and recalling that we have already evaluated the last term, gives

$$\int_{y=y_1}^{\infty} a_n^2 y^{-2}\, dy \le \left(\frac{y_0}{y_1}\right)^{2s} \int_{y=y_0}^{\infty} a_n^2 y^{-2}\, dy.$$

Summing over the a_n's and b_n's now gives us the theorem.

12.4 Belyi Surfaces

We begin with the following

Definition 1. *A compact Riemann surface S is called a Belyi surface if it admits a holomorphic function $f : S \to S^2$ such that f has at most three critical values.*

After a Möbius transformation, we may assume that the three points are 0, 1, and ∞. If we set

$$S^O = S - f^{-1}\{0, 1, \infty\},$$

we may characterize Belyi surfaces in the following way:

Theorem 3. *S is a Belyi surface if and only if there is a finite set of points $\{z_1, \dots, z_n\}$ on S such that any one of the following conditions is fulfilled on $S^O = S - \{z_1, \dots, z_n\}$:*

1. *$S^O = \mathbb{H}^2/\Gamma, \Gamma$ a finite index subgroup of $PSL(2, \mathbb{Z})$.*
2. *There is a graph G and an orientation $\mathcal{O}$ on G such that*

$$S^O = S^O(G, \mathcal{O}).$$

3. *S^O carries a horocycle packing – that is, a system of closed horocycles $\{C_i\}$ about the cusps $\{z_i\}$ such that the C_i's have disjoint interiors, and the region exterior to all the horocycles consist of triangular regions.*

We note that neither the Belyi function f nor the oriented graph $(G, \mathcal{O})$ are determined by S. If f is such a function, then composing f with $z \mapsto z^n$ produces another Belyi function, which gives rise to a new graph $(G, \mathcal{O})$. Proof of Theorem 3:

Let us first show that (1) holds if and only if S is a Belyi surface. If S is Belyi, then $S - f^{-1}(\{0, 1, \infty\})$ is a covering space of $S^2 - \{0, 1, \infty\}$. But

$$S^2 - \{0, 1, \infty\} = \mathbb{H}/\Gamma_2,$$

where

$$\Gamma_2 = \left\{ \begin{pmatrix} a\ b \\ c\ d \end{pmatrix} \right\} \equiv \pm \begin{pmatrix} 1\ 0 \\ 0\ 1 \end{pmatrix} \pmod 2$$

Conversely, suppose that there is a finite set $\{z_1, \dots, z_2\}$ in S such that $S - \{z_1, \dots, z_2\} = \mathbb{H}^2/\Gamma$ for some $\Gamma \subset PSL(2, \mathbb{Z})$.

We recall the well-known fundamental domain F_0 for $PSL(2, \mathbb{Z})$ acting on $\mathbb{H}^2$ acting $\mathbb{H}^2$ – it is traditionally written as

$$F_0 = \{z \in \mathbb{H} : -1/2 \leq \Re(z) \leq 1/2, |z| \geq 1\}$$

but we will find it more convenient to use the fundamental domain

$$F_0 = \{z \in \mathbb{H} : 0 \leq \Re(z) \leq 1, |z| \geq 1\, |z - 1| \geq 1\}.$$

This fundamental domain is shown in Fig. 12.4. It is given by cutting the fundamental domain F_0 along the line $\Re(z) = 0$ and regluing the left-hand piece to the right-hand side.

If on F we now glue the left-hand side to the right-hand side, we obtain a surface S_0, which is topologically a once-punctured sphere, with two orbifold points of orders 2 and 3, respectively, corresponding to the point $i \sim i+1$ and

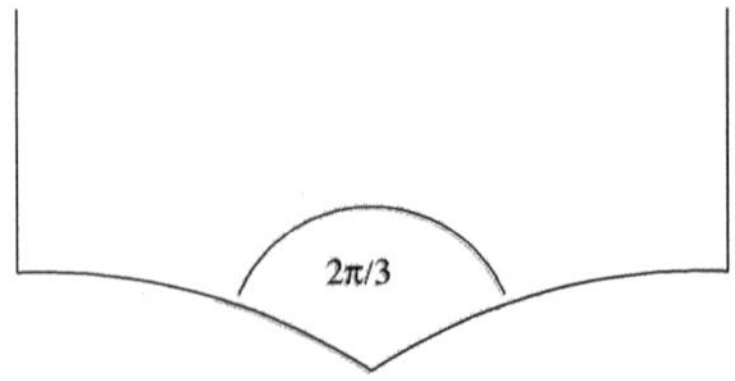

Fig. 12.4. The fundamental domain for $PSL(2,\mathbb{Z})$

$(1+i\sqrt{3})/2$, respectively. Removing the inverse images of these two points from S, we now get a covering

$$g : S - f^{-1}\left(i, \frac{1+i\sqrt{3}}{2}\right) \to S^2 - \{3.\text{points}\}$$

The points we remove from S are clearly removable singularities of g, so by filing them in, we get a branched cover $\tilde{g} : S \to S^2$ branched only over three points.

It is easy to see that (1) is equivalent to (2). Indeed, the oriented graph $(G, \mathcal{O})$ exhibits S^O as an orbifold covering of $\mathbb{H}^2/PSL(2,\mathbb{Z})$. To see that (1) implies (3), we may lift the horocycle on F joining i to $i+1$ to S^O. The corresponding system of horocycles has the desired properties. Conversely, suppose that S^O has the desired system of horocycles. Joining two horocycles of S^O by geodesics if the corresponding horocycles are tangent, we obtain a decomposition of S^O into ideal triangles. Each ideal triangle has three horocycles perpendicular to the edges and pairwise tangent. The only way this can happen is if the points of tangency are at the points i, $i+1$, and $(i+1)/2$. It follows that $S^O = S^O(G, \mathcal{O})$ for some $(G, \mathcal{O})$.

We will now show:

Theorem 4 (Belyi). *Let S be a closed Riemann surface. Suppose there exists a number field K and a holomorphic function*

$$f : S \to S^2$$

whose critical values lie in $K \cup \{\infty\}$.

Then S is a Belyi surface.

Our proof is based on the proof in [4]. The converse to this theorem is also true, and proved by Belyi, but we will not need it.

Proof. : Let $\phi : S^2 \to S^2$ be a rational function. If f is a holomorphic function from S to S^2 , then the critical values of $\phi(f)$ are the image under ϕ of the critical values of f, together with the critical values of ϕ.

Suppose that $z_1, \ldots, z^n$ are n points on S^2 lying in $K \cup \{\infty\}$, and let k be the maximal degree of the z_i's, which we may assume is z_1. Then there is a polynomial P of degree k with integer coeffcients such that $P(z_1) = 0$.

P does not raise the degree of any of the points z_i, and sends z_1 to something of degree 1, namely the point 0. Furthermore, P introduces critical points of degree $\leq k-1$, namely the solutions of $P'(z) = 0$. Hence the number of critical points of degree $\geq n$ reduces by at least one.

Arguing inductively, we may reduce to the case where all the z_i's have degree 1, that is they are rational numbers. After adding and multiplying by rational constants, we may assume that the critical values include 0, $1, \infty$, and at least one value between 0 and 1, which we may then write as $\alpha/(\alpha+\beta)$.

We now consider the map

$$P(z) z^{\alpha} (1-z)^{\beta}.$$

It sends 0 and 1 to 0, ∞ to ∞, and has critical points at (at most) 0, 1, and $\alpha/(\alpha+\beta)$. So the total number of critical values decreases by at least one. This concludes the proof.

As a simple consequence, we have

Corollary 1. *Let S be a Riemann surface. Then for every ε, there exists a surface S_ε within ε of S, such that*

$$S_\epsilon = S^C(G, \mathcal{O})$$

for some $(G, \mathcal{O})$.

Proof. According to Riemann–Roch, for any Riemann surface S there is a holoporphic function

$$f : S \to S^2.$$

We do not specify what metric we use to measure ϵ, because they are all equivalent. We could use any of the standard notions of distance in moduli space in the proof.

Take a small diffeomorphism of S^2 taking the critical points of f to points lying in some number field K, for instance the field $\mathbb{Q}[i]$, which is already dense in S^2. Then lift this conformal structure to S to obtain S_ϵ.

12.5 The Basic Construction

Let Γ be a three-regular graph. An orientation $\mathcal{O}$ on Γ is an assignment, for each vertex v of Γ, of a cyclic ordering of the edges coming out of each vertex. There is no "compatibility" requirement, so that a graph on n vertices will have in general 2^n orientation.

To the pair $(\Gamma, \mathcal{O})$, will assign a Riemann surface. Actually, we will assign two Riemann surfaces, $S^O(\Gamma, \mathcal{O})$ and $S^C(\Gamma, \mathcal{O})$. The surface $S^O(\Gamma, \mathcal{O})$ will be a finite area Riemann surface, while the surface $S^C(\Gamma, \mathcal{O})$ will be the conformal compactification of $S^O(\Gamma, \mathcal{O})$ obtained by filling in each cusp with a point.

We begin by considering the ideal hyperbolic triangle T (Fig. 12.5), which is the unique triangle with verticles at $0, 1, \infty$. We will mark some points on T – we will marks at the points i, $i+1$, and $(i+1)/2$, which can be thought of as the midpoints of the sides. We then join these points by horocycles. We then consider the point $(1+i\sqrt{3})/2$, and draw the geodesics joining this point to the three midpoints. To finish things up, we will draw a "traffic pattern" on T, showing the cyclic ordering corresponding to always turning left.

Here is what it looks like:

To the pair $(\Gamma, \mathcal{O})$ we construct the surface $S^O(\Gamma, \mathcal{O})$ in the following way: at each vertex, we place a copy of T , so that the traffic pattern on T matches up with the orientation at the vertex. Whenever two vertices are joined by an edge, we glue the corresponding triangles together so that the tick marks are glued together and the orientations match up (Fig. 12.6). This describes a unique gluing procedure.

We remark that the horocycle pieces on each T glue together to give closed horocycles about a cusp. Indeed, each cusp on the surface $S^O(\Gamma, \mathcal{O})$ corresponds to a path on the graph such that each time you arrive at a vertex, you turn left. We will call such paths *left-hand-turn* paths, or LHT paths for short.

It is easy to see that each surface $S^C(\Gamma, \mathcal{O})$ is a Belyi surface, and conversely, each Belyi surface arises this way. The oriented graph $(\Gamma, \mathcal{O})$ describes

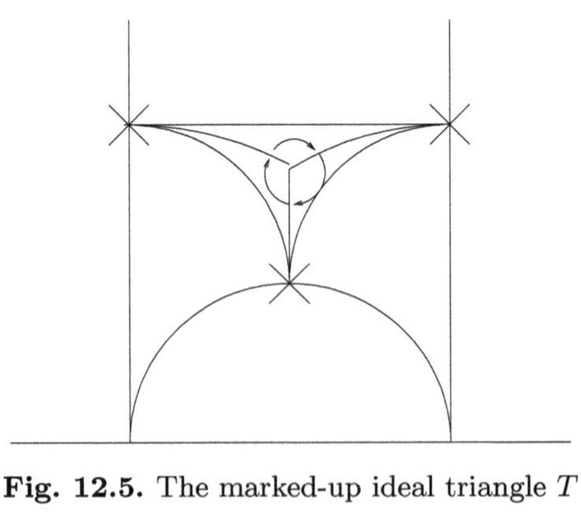

Fig. 12.5. The marked-up ideal triangle T

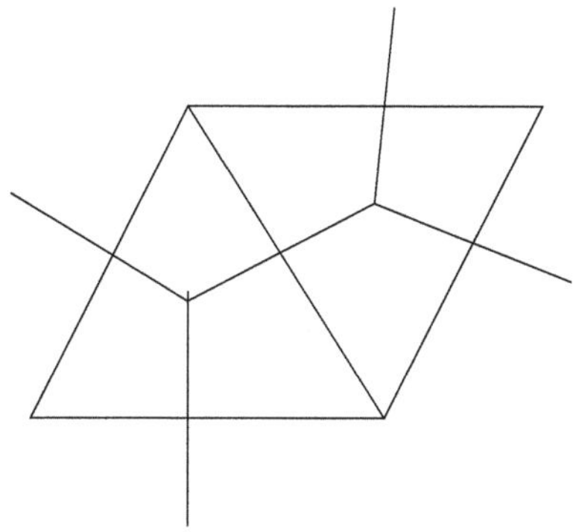

Fig. 12.6. The graph on two vertices

$S^O(\Gamma, \mathcal{O})$ as a covering space of $\mathbb{H}^2/PSL(2,\mathbb{Z})$, with the vertex being an orbifold point of order 3.

We remark that the surface so constructed depends in a very heavy way on the orientation. The easiest way to see this is by seeing on the genus. According to Gauss–Bonnet, the genus of $S^C(\Gamma, \mathcal{O})$ can be computed by

$$\text{genus}\,(S^C(\Gamma, \mathcal{O})) = 1 + \frac{n - 2LHT}{4}.$$

This gives us as an amusing sidelight that the number of LHT paths must have the same parity. Let us take some simple examples of this construction.

We first take the simplest graph, the three-regular graph on two vertices with no loops or double edges. It has two possible orientations.

Let us now build the surface $S^C(\Gamma, \mathcal{O})$. We begin by gluing two triangles together, as shown below:

In order to glue the remaining two pairs of sides, we need to use the orientation. With the first orientation, the left-hand side is glued to the top, while the right-hand side is glued to the bottom. The resulting surface is a sphere with three punctures, so its compactification must be the sphere.

With the second orientation, the left-hand side is glued to the right-hand side, while the top is glued to the bottom. We obtain in this way a torus with one cusp.

Which torus is it? The compactification process tells us that there is a unique way of assigning angles to the corners, so that first of all the angle sum around a cusp is 2π, and second so that the tick marks continue to be glued to tick marks. It is easy to see that assigning an angle of $\pi/3$ to each corner fulfills this requirement, because in an equilateral triangle the conformal midpoint is also the geometric midpoint.

Thus the surface $S^C(\Gamma, \mathcal{O})$ *equilateral torus*, obtained by gluing opposite sides of a parallelogram, is obtained by gluing two equilateral tori together.

Now let us consider the complete graph on four vertices, also known as the tetrahedron (Fig. 12.7). It has three essentially distinct orientations — you

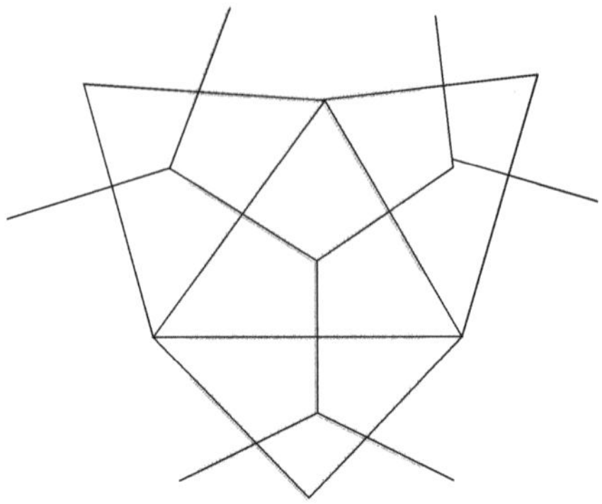

Fig. 12.7. Building on the tetrahedron

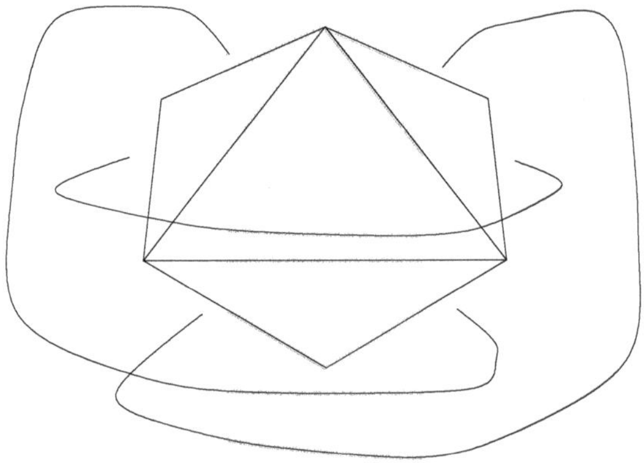

Fig. 12.8. The tetrahedron with one orientation reversed

may take the standard orientation and then reverse the orientation on zero, one, or two vertices (Fig. 12.8).

We begin by drawing four triangles glued together, with one in the center, as shown:

The usual orientation then tells us to glue each side to the side adjoining it so that it does not lie on the same triangle. As before, it is easy to see that one obtains in this way a sphere, this time with four singular points.

Now let us reverse the orientation on the center triangle, or what amounts to the same thing, keep the orientation on the center triangle, and reverse it in the other three.

Now we have that each side is glued to the opposite side. It is clear that there are now two cusps rather than four, so the surface is a torus.

One way to compute which torus it is, is as follows: from symmetry considerations, the central triangle must be an equilateral triangle. If we choose the remaining triangles to be isosceles, then the condition on the midpoints matching up is realized. This is because the conformal center of the base is again the geometric center, and, while it is not the case that the conformal center of the sides is the geometric center, it is the same for all sides, so whatever point it is, the points match up. In order to get the angle sums of the cusps to match up, the correct choice is $2\pi/3$ for the top angle, and $\pi/6$ for the bottom angles. When this is done, the picture looks as follows, and we have a regular hexagon with opposite sides glued together:

It is a nice exercise to cut and paste this shape together to see that it represents the equilateral torus. A faster way of seeing this is to notice the obvious order 3 symmetry (also clear for the graph) and observe that the equilateral torus is the only torus with an order 3 symmetry.

Now we consider the case where we reverse the orientation at two vertice (Fig. 12.9). Here there are two LHT paths, one passing through all four vertices and the other passing through all four vertices twice.

It is convenient to rearrange this picture with the four vertices from the four different triangles meeting at the central point, and with opposite sides identified. Using the argument with isosceles triangles we gave above, it is clear that the correct choice is for each triangle to be a $(\pi/4, \pi/4, \pi/2)$ isoseles triangle. From this it is easy to identify which torus this is – it is the *square torus*. This can also be seen directly from symmetry considerations.

As a final exercise, which we will not work out completely, the reader is invited to take the three-regular graph, which is the skeleton of a cube (Fig. 12.10). When one takes the usual orientation, then one again has a sphere, this time with six punctures. In Fig. 12.10, we have written down a number of possible ways of reversing orientations, which we denote by putting a circle around the corresponding vertex. Which surfaces do these represent?

In the first two examples, it is clear that the resulting surface is a torus, because there are four LHT paths – all of length 6 in the first example, and of lengths 4, 4, 6, and 10 in the second. In the third and fourth examples, we

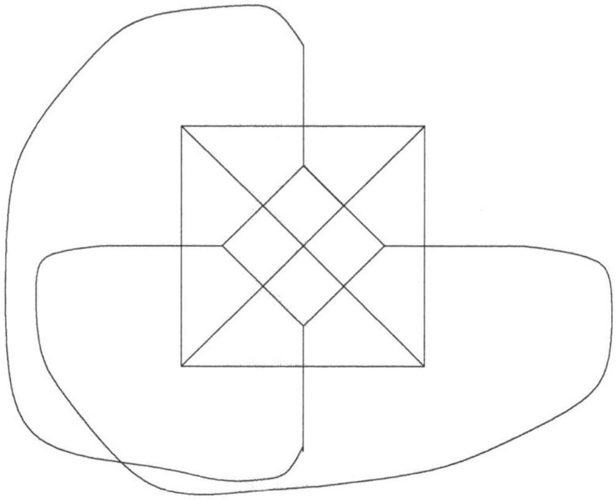

Fig. 12.9. The tetrahedron with two orientations reversed

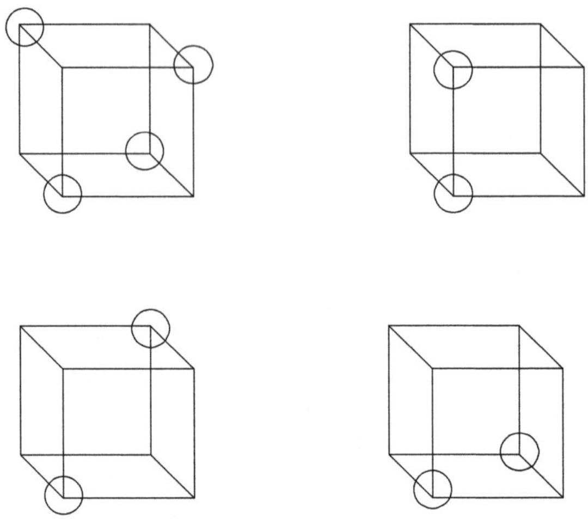

Fig. 12.10. Various options for orientations on the cube

got surfaces of genus two, because there are two LHT paths (two of length 12 in the third example, one of length 20 and one of length 4 in the fourth).

It is difficult to decide which surfaces are represented by the last two orientations, mostly because it is difficult to find names for surfaces of genus two.

12.6 The Ahlfors–Schwarz Lemma

The Ahlfors–Schwarz lemma should be familiar to students of complex analysis. Denoting by $\mathbb{D}$ the unit disk

$$\mathbb{D} = \{z \in \mathbb{C} : |z| < 1\},$$

it states

Lemma 1 (Schwarz). *Let $f : \mathbb{D} \to \mathbb{D}$ be a holomorphic map, which takes 0 to 0.*

Then

$$|f'(z)| \leq 1$$

and

$$|f(z)| \leq |z|, \quad z \neq 0,$$

with equality in either of the two inequalities at any point if and only if

$$f(z) = e^{i\theta}$$

for some θ.

Proof. Consider the function

$$g(z) = \begin{cases} \frac{f(z)}{z}, & z \neq 0 \\ f'(0), & z = 0. \end{cases}$$

On $|z| = r < 1$, we have

$$|g(z)| = \frac{|f(z)|}{|z|} \leq \frac{1}{r}.$$

Hence $|g(z)| \leq 1/r$ for all $|z| \leq r$. Letting $r \to 1$ complete the proof.

It was observed by Pick that the Schwarz lemma admits a geometric interpretation: if $\mathrm{d}s^2_{\mathbb{H}}$ denotes the hyperbolic metric on $\mathbb{D}$,

$$\mathrm{d}s^2 = \frac{4(\mathrm{d}x^2 + \mathrm{d}y^2)}{(1 - |z|^2)^2},$$

then any holomorphic map $\mathbb{D} \to \mathbb{D}$ is distance nonincreasing, with the distance between two points preserved if and only if f is an isometry, that is to say a Möbius transformation preserving $\mathbb{D}$.

The proof follows from what we have already done by first observing that the Möbius transformations preserving the disk are isometries of the metric $\mathrm{d}s^2_{\mathbb{H}}$, and secondly by composing f with such a Möbius transformation so that it sends 0 to 0.

Ahlfors observed that the argument could be made even more geometric by introducing the Gauss curvature κ of a metric. If the metric $\mathrm{d}s^2$ is given in conformal coordinates by $\mathrm{d}s$

$$\mathrm{d}s^2 = \lambda^2(z)[\mathrm{d}x^2 + \mathrm{d}y^2],$$

then

$$\kappa(\mathrm{d}s^2) = \frac{-\Delta(\log(\lambda))}{\lambda^2},$$

where, since we are currently in "analyst's mode,"

$$\Delta(f) = f_{xx} + f_{yy}.$$

The proof is an elementary calculation using Christoffel symbols.

We will give a number of different versions of the Ahlfors–Schwarz lemma. The first is due to Scott Wolpert:

Lemma 2. *Let S_1 and S_2 be two compact Riemann surfaces, with metrics $\mathrm{d}s_1^2$ and $\mathrm{d}s_2^2$ respectively, and*

$$f : S_1 \to S_2$$

a holomorphic map.

Suppose that $\kappa_2(f(z)) < \kappa_1(z) < 0$. Then f is distance decreasing.

Proof. : Let us write the "pullback metric" of $\mathrm{d}s_2^2$ by

$$f^*\mathrm{d}s_2^2 = g^2(z)\mathrm{d}s_1^2,$$

where $g(z)$ is a real function, which will be zero at the critical points of f. The fact that f is a holomorphic function is expressed here by the fact that $f^*(\mathrm{d}s_2^2)$ is conformal to $\mathrm{d}s_1^2$. f will be distance decreasing provided that $g < 1$.

The idea of the proof is as follows: let z_0 be a point at which g attains its maximum value. Such a point has to exist, since S_1 is compact. If z_0 is a branch point, then $g(z) \equiv 0$, and f is a constant (clearly distance decreasing). Otherwise, we may choose local coordinates about z_0, and use the same coordinates about $f(z_0)$.

Writing in these coordinates

$$\mathrm{d}s_1^2 = \lambda_1^2(z)|\mathrm{d}z|^2$$

and

$$\mathrm{d}s_2^2 = \lambda_2(z)|\mathrm{d}z|^2,$$

we clearly have that

$$g(z) = \frac{\lambda_2}{\lambda_1},$$

while

$$\kappa_1 = -\frac{\Delta(\log(\lambda_1))}{\lambda_1^2},$$

$$\kappa_2 = -\frac{\Delta(\log(\lambda_2))}{\lambda_2^2},$$

At z_0, we have

$$\Delta(\log(g)) \leq 0.$$

hence,

$$\Delta(\log(g)) = \Delta(\log(\lambda_2)) - \Delta(\log(\lambda_1)) = -\kappa_2\lambda_2^2 + \kappa_1\lambda_1^2 \leq 0$$

or

$$g^2 = \frac{\lambda_2^2}{\lambda_1^2} \leq \frac{(-\kappa_1)}{(-\kappa_2)} < 1$$

Hence $g < 1$, and the proof is complete.

Here is Alfors original version

Lemma 3. *Let* $\mathrm{d}s_1^2 = \mathrm{d}s_{\mathbb{H}}^2$ *and* $\mathrm{d}s_2$ *be two conformally equivalent metrics on* $\mathbb{D}$, *with* $\kappa_2 \leq -1 = \kappa_1$. *Then any map* $f : \mathbb{D} \to \mathbb{D}$ *is distance nonincreasing from* $\mathrm{d}s_1$ *to* $\mathrm{d}s_2$.

The proof will be a combination of the two arguments above. We again look at the function $g(z)$, and at an interior maximum z_0 it is clear that $g(z_0) \leq 1$, by the same curvature calculation as before. But how do we produce an interior maximum?

The idea is to look at the family of functions

$$f_r(z) = f(rz)$$

for $r < 1$, and the corresponding functions g_r. We may take the disk $\{z : |z| < r\}$ as coordinates to see that g_r goes to 0 at the boundary. This is because in this coordinate, clearly $\lambda_1(z) \to \infty$ as $|z| \to r$, while λ_2 remains finite.

Hence, if $g_r(z)$ is not identically 0, it must have an interior maximum, so $g_r(z) \leq 1$ everywhere. But clearly $g_r(z) \to g(z)$ as $r \to 1$, so we conclude that $g(z) \leq 1$ everywhere. This gives the lemma.

Here is the version we will be using.

Theorem 5. *Let $f : S_1 \to S_2$ be a holomorphic map between two (not necessarily compact) Riemann surfaces S_1 and S_2, and let $\mathrm{d}s_1$ and $\mathrm{d}s_2$ be metrics on S_1 and S_2 respectively.*

Suppose that the metric $\mathrm{d}s_1$ is complete, and that

$$\sup_{z \in S_2} \kappa_2(z) < \inf_{z \in S_1} \kappa_1(z) \leq \sup_{z \in S_1} \kappa_2(z) < 0.$$

Then f is distance nonincreasing.

Proof. After passing to the universal coverings, we may assume that S_1 and S_2 are both $\mathbb{D}$.

We now write $f_r(z) = f(rz)$ as before. The argument is exactly the same as in Ahlfors' argument, except at one small point. As before, the function g_r must go to zero at the boundary, from completeness of $\mathrm{d}s_1$. The only difference is that instead of using the pointwise estimate $\kappa_2 \leq -1$, we have to make do with comparisons of curvature at different points. Thus we must replace pointwise curvature estimates with sup and inf estimates.

This concludes the proof.

We note at this point that the map f no longer plays much of a role. There is no loss in assuming that instead of two Riemann surfaces and a map between them, we deal with one Riemann surface and two conformally equivalent metrics on it. The role of the function can be replaced by allowing the second metric to degenerate at some points. Here is our final version of the Ahlfors–Schwarz lemma. We present it as a corollary to the previous version. We would like to think of this as the "geometer's version" of Ahlfors–Schwarz, because it gives a nice, clean geometric statement, but from the analyst's point of view it may miss a lot that is covered by the lemma.

Corollary 2. *Let S be a Riemann surface with two complete metrics $\mathrm{d}s_1$ and $\mathrm{d}s_2$, which are conformally equivalent.*

Suppose that there exist constants C_1 and C_2 such that

$$C_1 \sup(\kappa_1) \inf(\kappa_2) \leq \sup(\kappa_2) \leq C_2 \inf(\kappa_1) \leq C_2 \sup(\kappa_1) < 0.$$

Then

$$C_2 \mathrm{d}s_2^2 \leq \mathrm{d}s_1^2 \leq C_1 \mathrm{d}s_2^2$$

Proof. After multiplying ds_1 by constants, we may apply the previous lemma. It is clear that after choosing the constants appropriately, we may change the role of $\mathrm{d}s_1$ and $\mathrm{d}s_2$. This completes the argument.

We like to paraphrase the corollary in the following way: "curvature close and negative implies metric close."

12.7 Large Cusps

In this section, we consider the following problem: let S^O be a noncompact finite-area Riemann surface, and let S^C be its conformal compactification. To what extent are the hyperbolic metrics on S^O and S^C related?

It is not too diffcult to see that there need be no relationship in general. For instance, if S^C is a sphere or a torus, and S^O is S^C with several points removed, then S^O will carry a hyperbolic metric, while S^C will not. Even if both S^O and S^C carry hyperbolic metrics, they will look quite different – S^C is compact, while S^O is not. However, one gets the feeling that for many geometric quantities arising in spectral geometry, the geometry does not really see things that take place far away on small regions. In particular, it should not make much difference to the body of the surface if a cusp is filled in or not.

In this section, we see how to realize this feeling. The key notion is the notion of *large cusps*:

Definition 2. *1. A cusp on S^O is of length $\geq L$ if there is a closed horocycle about the cusp whose length is at least L.*
2. The surface S^O has cusps of length $\geq L$ if there is a collection of horocycles with disjoint interiors, one enclosing each cusp, such that each horocycle has length at least L.

The main result of this section is then:

Theorem 6. *For every ε, there exists $L = L(\varepsilon)$ such that, if S^O is a finite-area hyperbolic Riemann surface with cusps of length $\geq L$, then outside standard cusp neighborhood on S^O and S^C, the hyperbolic metrics $\mathrm{d}s_O^2$ and $\mathrm{d}s_C^2$ satisfy*

$$\frac{1}{1+\varepsilon} \mathrm{d}s_0^2 \geq \mathrm{d}s_C^2 \geq (1+\varepsilon) \mathrm{d}s_0^2.$$

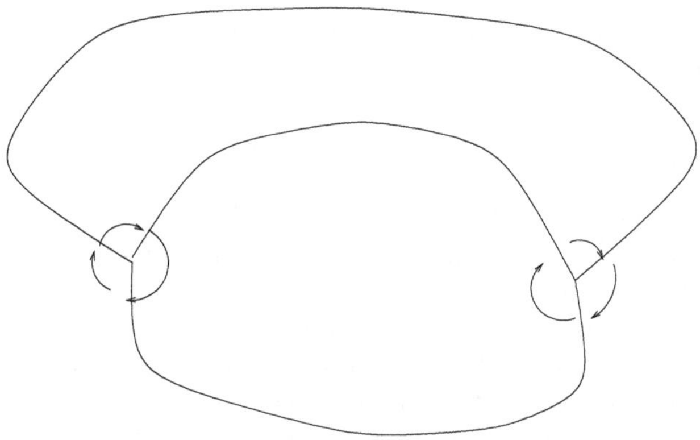

Fig. 12.11. A simple example

We remark that part of the statement of the theorem is that, in the presence of large cusps, S^C has a hyperbolic metric. One can see from Gauss–Bonnet that this will be the case if $L > 2\pi$. In effect, filling in a cusp takes away 2π from the Gauss–Bonnet integrand. A horocycle of length L binds a region of area L. So if all the horocycles have length $\geq L > 2\pi$, there is enough area left over so that S^C has a negative Euler characteristic, and hence a hyperbolic metric.

This argument is reasonably sharp. Let $(G, \mathcal{O})$ be the graph given in Fig. 12.11. Then, as we have seen in Sect. 12.5, $S^0(G, \mathcal{O})$ is the equilateral torus with one puncture, and the standard horocycle on $S^O(G, \mathcal{O})$ has length 6, which is just a little bit less than 2π. But $S^C(G, \mathcal{O})$ is a torus, and hence doesn't carry a hyperbolic metric.

Theorem 6.1 has a converse:

Theorem 7. *For every ε, there exists $R = R(\varepsilon)$ with the following property:*

Let S^C be a compact hyperbolic surface, and $z_1, \dots, z_k$ points on S^C such that the injectivity radii about the z^i's are at least R, and such that the balls $B(z_i, R)$ are disjoint.

Let $^O = S^C - \{z_1, \dots, z_k\}$, with its hyperbolic metric.

Then S^o has cusps of length $\geq \sinh(1/(1+\varepsilon)R)$, and outside of cusp neighborhoods, we have

$$\frac{1}{1+\varepsilon}\mathrm{ds}_0^2 \geq \mathrm{ds}_C^2 \geq (1+\varepsilon)\mathrm{ds}_0^2.$$

The strategy of the proof of Theorem 6.1 is as follows: we will consider two conformally equivalent metrics on S^C. The first metric will be the hyperbolic metric ds_C^2 on S^C. The second metric will be of the form

$$\widetilde{\mathrm{ds}_O^2} = f^2(z)\mathrm{ds}_O^2,$$

where f will be a function that is equal to one outside of cusp neighborhoods. We will want to choose f so that:

1. The metric $\widetilde{ds^2_O}$ extends to be a smooth metric across the cusps;
2. The curvature $\kappa(\widetilde{ds^2_O})$ lies between the $-1/1+\varepsilon$ and $-(1+\varepsilon)$.

The Alfors–Schwarz lemma, in the Wolpert version, will then guarantee that the metrics ds^2_C and $\widetilde{ds^2_O}$ are close. This will then establish the theorem.

To prove Theorem 6.2, we will proceed in the same way, reversing the roles of S^O and S^C. For this, we will need the noncompact version of the Ahlfors–Schwarz theorem. We must choose f so that it gives us a complete metric near the points z_i, in addition to the curvature estimates.

To set up the basic calculation, let $\mathbb{D}$ denote the disk, and let $ds^2_{\mathbb{D}}$ be the hyperbolic metric on $\mathbb{D}$

$$ds^2_{\mathbb{D}} = \frac{4}{(1-|z|^2)^2}[dx^2 + dy^2].$$

We will also consider the hyperbolic metric ds^2_H on $\mathbb{D} - 0$. It is given by the formula

$$ds^2_H = \left(\frac{1}{-|z|\log(|z|)}\right)^2 [dx^2 + dy^2].$$

Note that the ratio

$$h^2 = \frac{ds^2_H}{ds^2_{\mathbb{D}}}$$

is given by

$$h = \frac{\frac{1}{-|z|\log(|z|)}}{\frac{2}{1-|z|^2}}.$$

It will be convenient to take geometric coordinates. Let $r(z)$ be the distance from 0 in the metric $ds^2_{\mathcal{D}}$. Then

$$r(z) = \int_0^{|z|} \frac{2}{1-x^2}dx = \int_0^{|z|}\left[\frac{1}{1+x} + \frac{1}{1-x}\right]dx = \log\left(\frac{1+|z|}{1-|z|}\right),$$

which gives the inverse map as

$$|z| = \tanh\left(\frac{r}{2}\right).$$

Writing

$$dx^2 + dy^2 = [d(|z|)^2 + |z|^2 d\theta^2],$$

we have

$$\frac{4}{(1-|z|^2)^2}[dx^2+dy^2] = \left[\frac{4}{(1-\tanh^2(r/2))^2}\right] \times \left[\left(\frac{1}{2\cosh^2(r/2)}\right)^2 dr^2 + \tanh^2(r/2)d\theta^2\right].$$

The coefficient of d/$theta^2$ is

$$\left[\left(\frac{2}{1-\tanh^2(r/2)}(\tanh(r/2))\right)\right]^2 = \left[\frac{2\sinh(r/2)\cosh(r/2)}{\cosh^2(r/2)-\sinh^2(r/2)}\right]^2 = \sinh^2(r/2)$$

while the coefficient of dr^2 is

$$\frac{2}{1-\tanh^2(r/2)}\frac{1}{2\cosh^2(r/2)} = 1,$$

so the metric reduces to

$$\mathrm{d}s^2_{\mathbb{D}} = \mathrm{d}r^2 + \sinh^2(r)\mathrm{d}\theta^2.$$

The function h may then be written as

$$h(r) = \frac{1}{\sinh(r)\log(\coth(r/2))},$$

so that the metric $\mathrm{d}s^2_{\mathbb{H}}$ is

$$\mathrm{d}s^2_{\mathbb{H}} = h(r)\left[\mathrm{d}r^2 + \sinh(r)\mathrm{d}\theta^2\right].$$

It will be convenient to have the formula for the curvature κ_g of a metric of the form $g^2\left[\mathrm{d}r^2 + (\sinh^2(r))\mathrm{d}\theta^2\right]$. It is given by

$$\kappa_g = -\left[\left(\frac{g'}{g}\right)' + 1 + \left(\frac{g'}{g}\right)\coth(r)\right].$$

Of course, it is in general difficult to decide what is really important in all these formulas. The main point of the curvature formula is that it involves two derivatives in g, and is close to -1 provided that g is close to 1 and its first and second derivatives are close to 0.

The main point about h is that it is a function for which, as $r \to \infty$, we have that $h(r)$ is close to 1, while h' and h'' are close to 0. This can be seen by some simple uses of L'Hospital's rule.

The idea of the proof of the theorem is now to find a function g, which is equal to 1 for small values of r, which is equal to h for large values of r, and for which the values of κ_g stay close to -1. It is easy that one can do this, provided that $g(r)$ agrees with 1 for large enough values of r so that $h(r)$ is close to 1 and the first two derivatives of h are close to 0. This then establishes Theorem 6.1. The proof of Theorem 6.2 goes exactly the same way, reversing the roles of $\mathrm{d}s^2_{\mathbb{H}}$ and $\mathrm{d}s^2_{\mathbb{D}}$.

12.8 The Spaghetti Model

We have seen how the Bollobas theory gives us a good picture of the spectral behavior of a general Riemann surface. A reasonable question to ask is whether there are other features of the geometry of Riemann surfaces which can be read off from this picture.

The key point here is to take seriously the additional structure afforded by the orientation of a graph. As we have seen, the orientation does not have a strong effect on the behavior of λ_1 (and indeed has no effect on the behavior of λ_1 of a graph), but it does have a very strong effect on the geometry of the surface.

Indeed, what is the difference between the two pictures of the Ramanujan graph $X^{2,3}$ given in the introduction? The point is that since $X^{2,3}$ is a homogeneous graph, it carries a natural orientation. It was the orientation that was responsible for unraveling the chaos of the first picture to obtain the order of the second picture.

In this section, we will describe the following two results:

Theorem 8. *Let $(\Gamma, \mathcal{O})$ be chosen randomly among oriented 3-regular graphs on n vertices. Then the expected value $E(\text{genus}(S^C(\Gamma, \mathcal{O}))$ satisfies*

$$(\text{const}) + (n/4) - (3/4)\log(n) \leq E(\text{genus}) \leq (\text{const}) + (n/4) - (1/2)\log(n)$$

Theorem 9. *If S is a Riemann surface, denote by $Emb(S)$ the area of the largest embedded ball in S. Then the expected value of $Emb(S^C(\Gamma, \mathcal{O}))$ satisfies*

$$E(Emb(S^C(\Gamma, \mathcal{O})) \geq (1/\pi)area(S).$$

Of course, the first theorem tells us more about our method of picking Riemann surfaces then it is about the surfaces themselves, but the second theorem tells us a fascinating fact about Riemann surfaces – the general Riemann surface has its geometry dominated by one very large embedded ball.

The idea of the proofs of these theorems is to translate them into what they are stating about LHT paths. According to our formula for the genus, the first theorem shows that the expected value of the number of LHT paths grows logarithmically in n. A reasonable question to ask is what one expects about the associated lengths of the left-hand turn paths. We claim that the second theorem shows us about the expected length of the longest LHT path:

Lemma 4. *Let $(\Gamma, \mathcal{O})$ be an oriented graph. For given LHT path C, let L_C be the length of this path. If $S^O(\Gamma, \mathcal{O})$ obeys the large cusp condition, then about the image of the corresponding cusp of $S^O(\Gamma, \mathcal{O})$ in $S^C(\Gamma, \mathcal{O})$, there exists an embedded ball of area $\sim L_C$.*

The proof is to apply the Ahlfors–Schwarz lemma. The horocycle neighborhood of C in $S^O(\Gamma, \mathcal{O})$ has area equal to the length of C, and this horocycle neighborhood goes over to an embedded ball in the compactification.

Before discussing the proofs of these theorems, we will say a few words concerning where they come from. If we pick an element $\sigma = \sigma_n$ randomly from the symmetric group $S(n)$ on n elements, we may ask to give the *cycle decomposition* of σ. This will of course determine σ up to conjugacy. We then have:

Theorem 10. *The cycle decomposition of a randomly picked element of* $S(n)$ *has the following properties:*

1. *The expected number of cycles is* $\sim \log(n)$.
2. *There is a constant* c *such that the expected length of the longest cycle is at least* $c \cdot n$.

I do not know to whom to attribute these results, but they have been actively studied and generalized in recent years. The precise value of c has been computed, and is something like $.62\ldots$, but we will present a simple argument showing that $c \geq 1/2$. The reason for doing this is that the simple argument will generalize to the setting of three-regular graphs, once we clarify what the right translation to this setting is. As a result, the constant $1/\pi$ appearing in Theorem 7.2 is certainly not sharp, and our belief is that it should probably be about twice as big, but already the simple estimate gives us a striking geometric fact.

We would like to leave the correct estimate of the constant on the hands of experts.

Theorem 7.3 can be seen easily from what has become known as *the spaghetti model.* Imagine n pieces of spaghetti, labeled from 1 to n, lined up so that each piece lies vertically. At each step of the process, one ties the bottom of one piece of spaghetti, chosen randomly, to the top of another piece of spaghetti, again chosen randomly. In this way, one creates an element of the symmetric group, and the number of components of spaghetti and their corresponding lengths correspond to the number of cycles and their respective lengths.

At the kth step of the process, the probability of forming a closed loop is exactly $1/(n-k+1)$, as is easily seen. Thus the expected number of closed loops at the end will be

$$1/n + 1/(n-1) + \cdots + 1 \sim \log(n).$$

Now let us modify the process in the following way: at the first step, the bottom piece of spaghetti is chosen to be the first one. At the kth step, we pick as the bottom piece the piece whose top piece was chosen at the $(k-1)$th step. We continue in this way until a closed loop is formed. What is the expected length of this loop?

Well, the probability of its length 1 is $1/n$. The probability of its length 2 is

$$[(n-1)/n][1/(n-1)] = (n+1)/2 \sim n/2.$$

In general, it is not hard to see that the probability of the process stopping at exactly k steps is exactly $1/n$. So the expected value is easily computed to be

$$(1/n)[1 + \cdots + n] = (n+1)/2 \sim n/2.$$

We remark that one could do a similar computation where one modifies the process by picking at random a free end, disregarding whether it is a top or a bottom, and joining it to another end, again disregarding whether it is a top or a bottom. One gets $\sim (1/2)\log(n)$ for the number of components, as before, but the expected length rises to $2n/3$, rather than $n/2$. A rather unexpected (at least to me) cancellation of terms arises, similar to the fact that we always got $1/n$, which makes this calculation elementary.

Note that in either case, we have not calculated the expected length of the *longest* loop, but rather the expected length of a loop starting from some given point. While we expect that this point will tend to lie on a longer rather than a shorter loop, the expected value of the longest path will be greater. This gives the improvement from $1/2$ to $.62\ldots$, but we prefer the ease of computation to the sharp constant.

We now consider how to translate this to the estimation of left-hand turn paths on three-regular oriented graphs. The starting point is to modify the spaghetti model in the following way: instead of putting down n pieces of spaghetti, we put down n pieces each of which is a vertex with three half-edge (Fig. 12.12):

We will find it convenient to draw on these pieces the corresponding pieces of LHT paths.

The process of picking up an end and gluing it to another end is exactly the Bollobas model, modified by labeling the balls so as to get a cyclic ordering.

We now want to make the same calculation we did before. When we pick up an end, what are the probabilities of forming a closed LHT path when we pick another end?

We may make the same calculation as before: look at the end we have picked up, and follow the two LHT path segments leading from it, until you arrive at two endpoints (which may be the same point). It would therefore appear that the expected value of the number of closed LHT paths produced would be 2 divided by the number of remaining edges, which in turn would give an estimate of $\sim \log(n)$ for the expected number of LHT paths.

The problem with this argument arises in the following picture:

If one LHT path starting at an end wraps around and finishes at the same end, then we call this end a *bottleneck* (Fig. 12.13). If two bottlenecks are joined together, then a closed LHT path is formed. Thus, the probability of forming a closed

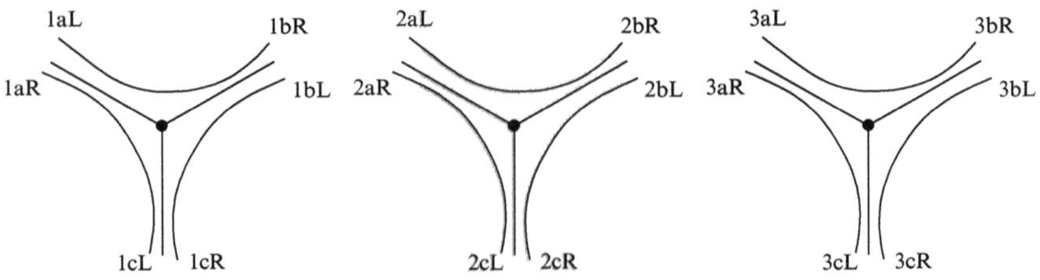

Fig. 12.12. Modified spaghetti

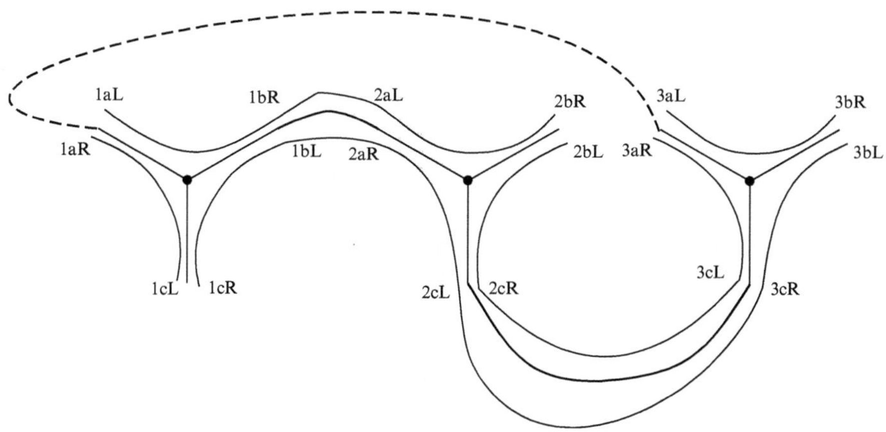

Fig. 12.13. Bottlenecks

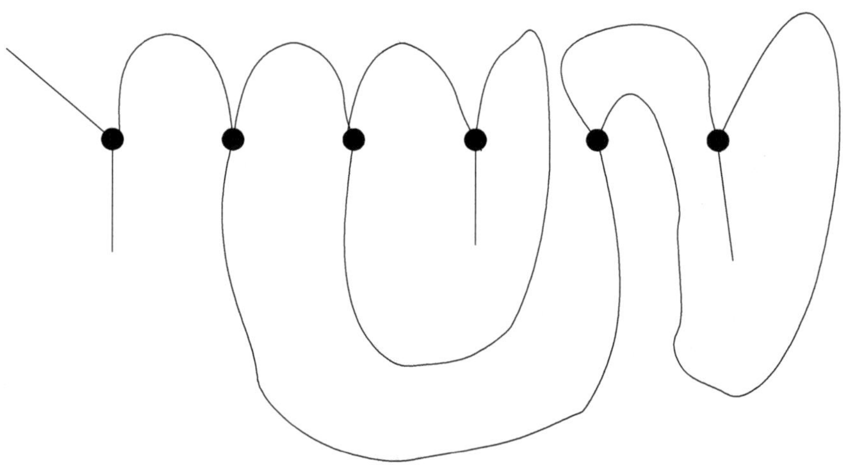

Fig. 12.14. An additional complication

LHT path is affected by how many bottlenecks are present, invalidating the previous argument.

We resolve this issue in the following way: when a bottleneck is formed, we count it as half an LHT path. We then only count closed LHT paths that are formed by gluing together two ends that are not bottlenecks. In this way, the previous argument providing a logarithmic bound remains intact, at the expense of raising the constant from 1 to 3/2. This is clearly an overestimate, since a bottleneck may be destroyed before it is joined to another bottleneck.

This establishes Theorem 7.1.

The argument to establish Theorem 7.2 is similar, but more complicated (Fig. 12.14). Again the problem is that we have to worry about the creation of bottle necks. There are also some more problems we have to worry about as well, as evidenced in the following picture:

But the heart of the idea is to use the result of Theorem 7.1 to see that these problems only arise late in the game, when the (logarithmically many) bottlenecks may be large compared to the number of remaining edges. But if we have gotten to that late in the game, then the LHT path is already quite long.

The constant $1/\pi$ should be thought of as $(1/3) \cdot (3\pi)$, where the second term arises because each fundamental domain has area $\pi/3$. The term $1/3$ should be thought of as the $2/3$ arising in the spaghetti model argument, divided by 2 since each edge has two adjoining LHT segments.

The constant is not sharp for a variety of reasons: first of all, because we are measuring the expected length starting from a given LHT segment, rather than the expected length of the longest LHT path. Second, when we are counting the length of the LHT path we are building, we assume that at each step the length increases by one this is what happens in the spaghetti model. But in fact, at each step we increase the length not just of the LHT path we are measuring, but also another LHT path. So as a point of fact the length of the LHT path we are building is really increasing much faster than we are counting.

This completes the sketch of the proof of Theorem 7.2.

12.9 An Annotated Bibliography

As mentioned in the introduction, the work discussed in these notes is spread out over a large number of papers. This happened because my thinking about this topic underwent a rather long development, during which different facets of the picture emerged. In what follows, I give an annotated guide to the papers I have written on the subject, together with coauthors. I have decided not to give a comprehensive bibliography on the subject here. All of these papers are available at my website, http://www.math.technion.ac.il/ rbrooks, except when they are old.

- [1]. contains an announcement of the results contained in [2], [3], and early versions of [4].
- [5]. This paper gives the theorem connecting the bottom of the spectrum of a covering and the Cheeger constant of the corresponding graph.
- [6]. This paper gives the compactification techniques using the Ahlfors–Schwarz lemma. It also gives a presentation of the Platonic graphs and their corresponding surfaces.
- [4]. This paper studies the process of building a surface at random by choosing a three-regular graph at random. This paper has gone through a number of different versions, the latest version includes results concerning the expected genus and the expected largest embedded ball of a randomly chosen Riemann surface.

- [2]. This paper shows how to construct Riemann surface with large first eigenvalue of arbitrary genus. It also contains growth estimates for eigenvalues in cusps as well as other generally useful techniques involving compactification and the behavior of eigenvalues.
- [7]. Contains a number of versions of the Ahlfors–Schwarz lemma, including the version needed in [2].
- [3]. Features graph-theoretic techniques for building Belyi surfaces with various nice properties.
- [8]. This contains the theorem relating Cheeger constants of graphs to the behavior of the first eigenvalue under coverings. (Too old to be available at my web site.)
- [9]. A survey article discussing the construction of building surfaces from three-regular graphs.

In addition to these papers, I would like to mention the M.Sc. thesis of my student Dan Mangoubi, "Riemann Surfaces and three-Regular Graphs," available from my web site. In addition to giving a good overview of the subject, it contains interesting quantitative results elaborating on the qualitative theory of [6]. There are also two papers in preparation: my paper with Andrzej Zuk on Cheeger constants of graphs and surfaces, and my paper with Mikhail Monastyrsky, which generalizes the theory from three-regular graphs to k-regular graphs.

References

1. R. Brooks, E. Makover, ERA-AMS **5** 76–81 (1999)
2. R. Brooks, E. Makover, J. d'Anal. **83** 243–258 (2001)
3. R. Brooks, E. Makover, Sodin et. al (eds.), *Entire functions in modern analysis*, IMCP **15** 37–46 (2002)
4. R. Brooks, E. Makover, Preprint, Department of Mathematics, Technion (1997)
5. R. Brooks, Comm. Math. Helv. **56** 581–596 (1981)
6. R. Brooks, Comm. Math. Helv. **74** 156–170 (1999)
7. R. Brooks, Brooks and Sodin (eds.), Lectures in Memory of Lars Ahlfors, IMCP **14**, 31–39
8. R. Brooks, J. Diff. Geom. **23** 97–107 (1986)
9. R. Brooks, Picardello and Woess (eds.), *Random walks and discrete potential theory*, (Camb. Univ. Press, 1999), pp. 85–103

Index

MIX
Papier aus verantwortungsvollen Quellen
Paper from responsible sources
FSC® C105338

If you have any concerns about our products,
you can contact us on
ProductSafety@springernature.com

In case Publisher is established outside the EU,
the EU authorized representative is:
Springer Nature Customer Service Center GmbH
Europaplatz 3, 69115 Heidelberg, Germany

Printed by Libri Plureos GmbH
in Hamburg, Germany